W9-DAC-290

Selection Bias and Covariate Imbalances in Randomized Clinical Trials

STATISTICS IN PRACTICE

Advisory Editor

Stephen Senn
University College London, UK

Founding Editor

Vic Barnett
Nottingham Trent University, UK

Statistics in Practice is an important international series of texts, which provide detailed coverage of statistical concepts, methods and worked case studies in specific fields of investigation and study.

With sound motivation and many worked practical examples, the books show in down-to-earth terms how to select and use an appropriate range of statistical techniques in a particular practical field within each title's special topic area.

The books provide statistical support for professionals and research workers across a range of employment fields and research environments. Subject areas covered include medicine and pharmaceutics; industry, finance and commerce; public services; the earth and environmental sciences, and so on.

The books also provide support to students studying statistical courses applied to the above areas. The demand for graduates to be equipped for the work environment has led to such courses becoming increasingly prevalent at universities and colleges.

It is our aim to present judiciously chosen and well-written workbooks to meet everyday practical needs. The feedback of views from readers will be most valuable to monitor the success of this aim.

A complete list of titles in this series appears at the end of the volume.

Selection Bias and Covariate Imbalances in Randomized Clinical Trials

Vance W. Berger
University of Maryland, Baltimore County, USA

John Wiley & Sons, Ltd

Other Wiley Editorial Offices

John Wiley & Sons Inc., 111 River Street, Hoboken, NJ 07030, USA

Jossey-Bass, 989 Market Street, San Francisco, CA 94103-1741, USA

Wiley-VCH Verlag GmbH, Boschstr. 12, D-69469 Weinheim, Germany

John Wiley & Sons Australia Ltd, 33 Park Road, Milton, Queensland 4064, Australia

John Wiley & Sons (Asia) Pte Ltd, 2 Clementi Loop #02-01, Jin Xing Distripark,
Singapore 129809

John Wiley & Sons Canada Ltd, 22 Worcester Road, Etobicoke, Ontario, Canada M9W 1L1

Wiley also publishes its books in a variety of electronic formats. Some content that appears
in print may not be available in electronic books.

Library of Congress Cataloguing-in-Publication Data

Berger, Vance.
    Selection bias and covariate imbalances in clinical trials / Vance W. Berger.
       p. cm.
    ISBN-13 978-0-470-86362-6
    ISBN-10 0-470-86362-5
    1. Clinical trails–Statistical methods–Evaluation.   2. Ranking and
selection (Statistics)–Evaluation.   3. Sampling (Statistics)–Evaluation.
I. Title.

    R853 .C55B47   2005
    610'.72'4—dc22   2004026799

British Library Cataloguing in Publication Data

A catalogue record for this book is available from the British Library

ISBN-13 978-0-470-86362-6 (HB)
ISBN-10 0-470-86362-5 (HB)

Typeset in 11/13pt Photina by TechBooks, New Delhi, India
Printed and bound in Great Britain by TJ International Ltd, Padstow, Cornwall
This book is printed on acid-free paper responsibly manufactured from sustainable forestry
in which at least two trees are planted for each one used for paper production.

Contents

Preface   ix

Part I: Is There a Problem with Reliability in Medical Studies?   1

1   An Evolution of Comparative Methodology   **3**

1.1   Single-subject studies   3
1.2   Case series and cohort studies   4
1.3   Historical controls   5
1.4   Parallel control groups   6
1.5   Matched studies   6
1.6   Randomization   9
1.7   Advance randomization   11
1.8   Allocation concealment   12
1.9   Residual selection bias   14

2   Susceptibility of Randomized Trials to Subversion and Selection Bias   **17**

2.1   Can randomized trials be subverted?   18
2.2   If randomized trials are subverted, do they cease to be randomized trials?   22
2.3   What is masking?   25
2.4   What is allocation concealment?   27
2.5   A double standard   28
2.6   What if allocation concealment could be ensured?   29

3   Evidence of Selection Bias in Randomized Trials   **37**

3.1   The burden of proof regarding the existence of selection bias in randomized trials   38
3.2   Indirect population-level evidence that selection bias exists in randomized trials   41
3.3   Direct trial-level evidence that selection bias exists in randomized trials   42
     3.3.1   Heparin for myocardial infarction   43
     3.3.2   University Group Diabetes Program   43

3.3.3	Talc and mustine for pleural effusions	44
3.3.4	Tonsillectomy for recurrent throat infection in children	45
3.3.5	Oxytocin and amniotomy for induction of labor	45
3.3.6	Western Washington Intracoronary Streptokinase Trial	46
3.3.7	RSV immune globulin in infants and young children with respiratory syncytial virus	47
3.3.8	A trial to assess episiotomy	47
3.3.9	Canadian National Breast Cancer Screening Study	48
3.3.10	Surgical trial	49
3.3.11	Lifestyle Heart Trial	50
3.3.12	Coronary Artery Surgery Study	50
3.3.13	Etanercept for children with juvenile rheumatoid arthritis	51
3.3.14	Edinburgh Randomized Trial of Breast-Cancer Screening	52
3.3.15	Captopril Prevention Project	53
3.3.16	Göteborg (Swedish) Mammography Trial	53
3.3.17	HIP Mammography Trial	53
3.3.18	Hypertension Detection and Follow-Up Program	63
3.3.19	Randomized trial to prevent vertical transmission of HIV-1	63
3.3.20	Effectiveness trial of a diagnostic test	63
3.3.21	South African trial of high-dose chemotherapy for metastatic breast cancer	64
3.3.22	Randomized study of a culturally sensitive AIDS education program	65
3.3.23	Runaway Youth Study	65
3.3.24	Cluster randomized trial of palliative care	67
3.3.25	Randomized trial of methadone with or without heroin	72
3.3.26	Randomized NINDS trial of tissue plasminogen activator for acute ischemic stroke	74
3.3.27	Norwegian Timolol Trial	75
3.3.28	Laparoscopic versus open appendectomy	77
3.3.29	The Losartan Intervention for Endpoint Reduction in Hypertension (LIFE) Study	78
3.3.30	The Heart Outcomes Prevention Evaluation (HOPE) Study	79
3.4	In search of better evidence	80

4   Impact of Selection Bias in Randomized Trials   85

4.1	Quantifying the prediction of future allocations: balanced blocks	85
4.2	Quantifying prediction of future allocations: unbalanced blocks	90
4.3	Quantifying covariate imbalance resulting from selection bias	94
4.4	Quantifying the bias resulting from covariate imbalance	100

Part II: Actions to be Taken to Improve the Reliability of Medical Studies   103

5   Preventing Selection Bias in Randomized Trials   105

5.1	Minimizing the Impact of Selection Bias	106
5.2	Biased Selection of Investigators	106

5.3	Minimizing the prediction of future allocations	107
	5.3.1	The trade-off between selection bias and chronological bias	109
	5.3.2	Notation	111
	5.3.3	Varying the block sizes	113
	5.3.4	The maximal procedure	115
	5.3.5	Extensions	121

6 Detecting Selection Bias in Randomized Trials 123

6.1	Baseline Imbalances in Observed Covariates	124
6.2	Testing for selection bias without baseline analyses	129
6.3	The selection covariate	129
6.4	The role of the reverse propensity score in third-order residual selection bias	130
6.5	Using the reverse propensity score To Test for selection bias: the Berger–Exner test and graph	132
6.6	Using the screening log to test for selection bias	142
6.7	The Ivanova-Barrier-Berger (IBB) Detection Method	145
6.8	Interpreting negative tests of selection bias	147
6.9	When should one test for selection bias?	148
6.10	Who should test for selection bias?	155

7 Adjusting for Selection Bias in Randomized Trials 157

7.1	Methods proposed for addressing non-random baseline imbalances	159
7.2	Selection bias arising from a complete lack of allocation concealment	161
7.3	Selection bias arising from imperfect allocation concealment	161
	7.3.1	The RPS approach to adjusting for selection bias	162
	7.3.2	The Ivanova–Barrier–Berger (IBB) method for correcting selection bias	165

8 Managing Selection Bias in Randomized Trials 171

8.1	Action points during the design phase of the trial	171
8.2	Action points during the conduct of the trial	174
8.3	Action points during the analysis of trial data	178
8.4	Action points by party	182

References	187
Author Index	199
Subject Index	203

Preface

It stands to reason that important decisions, such as those involving the treatment of a patient or a group of patients, would be based on solid evidence, especially when one considers what the alternative is. Hence, evidence-based medicine has become a major theme of medical research and medical decision-making. A hierarchy of evidence has evolved to guide users of medical research, and randomized trials tend to gravitate towards the top of this hierarchy. In fact, it is not uncommon to hear the phrase 'randomized evidence' used to denote the strongest of all possible types of evidence.

It is far less common to find discussions of the ways in which variation among randomized trials themselves might make some better than others within the hierarchy of evidence. At the heart of the unspoken assumption that all randomized trials are created equal is the belief that all randomized trials are fair comparisons. That is, the validity of the comparison is guaranteed by virtue of the fact that the word 'randomized' appears in the title. The only question, then, is how well these internally valid randomized results generalize to the target population. That is, to what extent is there external validity? The selection bias most commonly associated with randomized trials is the one that might limit the external validity, because the results of randomized trials would tend to apply to only those patients willing to be randomized. This is a legitimate concern, but it is not the concern, or the type of selection bias, considered in this book.

Before one can even ask if the results of a randomized trial generalize, one needs to ask if these results are internally valid. Upon careful scrutiny it is fairly easy to deduce that internal validity, or a fair comparison, is in fact *not* guaranteed by even proper randomization (which, by the way, is also not guaranteed merely by virtue of the claim that the trial was randomized). The central theme of the book is the idea that even in a properly randomized trial confounding can be induced to create a covariate imbalance that leads to a type of selection bias that can compromise internal validity. That is, estimates of treatment effects can be biased, confidence intervals can be too narrow, and p-values can be artificially too low as a result of this selection bias. Moreover, nothing in these summary statistics themselves will allow a reader to discern that such selection bias has occurred, so one needs to look elsewhere to confirm the face validity of the findings.

This book deals with the aspects of selection bias that one would need to consider when designing a randomized trial or analyzing the data from a randomized trial or reviewing the results of a randomized trial. Part I (Chapters 1–4) serves as a statement of the problem. Specifically, Chapter 1 presents the various types of study designs and clarifies why randomized trials are at the top of the hierarchy of evidence. Chapter 2 presents the mechanisms by which upcoming treatment allocations can be predicted even in randomized trials, and how this foreknowledge of future allocations can be exploited to create the type of selection bias we consider. Chapter 3 presents the evidence that this type of selection bias has actually occurred, and is not simply a hypothetical concern. Chapter 4 presents the impact this selection bias can have on the results of randomized trials. It is suggested that Chapter 2 be read by all readers, because it sets the stage for the methods proposed in subsequent chapters for managing selection bias. Those readers who are studying selection bias as part of their training in clinical trial design and analysis might wish to also read Chapters 1, 3, and 4, but these chapters are optional on a first reading for those readers who have actual trials (and deadlines!) in mind when reading this book, and wish to apply the results as quickly as possible. These readers should skip ahead to Part II right after reading Chapter 2.

Part II tells the reader the steps that can be taken to manage selection bias in randomized trials. Chapter 5 deals with prevention by various means, but mostly by selecting a randomization technique that defies prediction of future allocations even when the past allocations are known. Of course, one could eliminate all prediction, but at the cost of enabling chronological bias, and so the trade-off between these two must be considered. A reader who is writing a protocol should read Chapter 5 right after Chapter 2. Chapter 6 deals with detecting selection bias by various methods, but mostly by the Berger–Exner test.

Readers who review randomized trials, such as regulatory agencies, journal editors, and funding agencies, might wish to read Chapter 6 right after Chapter 2. Chapter 7 deals with methods recently proposed for correcting for selection bias if it is found in a randomized trial. This would be important reading for any reader who has found even the appearance of selection bias in a randomized trial. Finally, Chapter 8 offers recommendations, and draws from previous chapters. A module on selection bias within the framework of a training program in clinical trial design and analysis might use just Chapters 2 and 8, with the other chapters serving as reference material. Chapter 8 is also an important chapter for researchers in trial design methodology, because it presents some open problems the solutions to which might help to improve future trial methodology.

This is not a mathematically challenging text. The reader will not encounter endless equations and derivations, and so there really is no mathematical prerequisite. In fact, this book is intended for a broad readership representing the full spectrum of disciplines contributing to the design, conduct, analysis, reporting, and review of randomized clinical trials. Perhaps the best prerequisites would be intellectual curiosity and perhaps actual studies to work on.

I wish to thank Stephen Senn for discussing these ideas with me, and for suggesting that I put these ideas into book form. In addition, it is his own writings on baseline imbalances in randomized trials that began my thinking about these ideas. I also thank the researchers who joined me in studying selection bias in randomized trials, including Derek Exner, Jeffrey Bears, Costas Christophi, Anastasia Ivanova, Maria Deloria-Knoll, and Sherri Weinstein. I thank Ken Schulz for being part of a successful workshop on this subject at the 2002 Society of Clinical Trials Meeting in Arlington, VA, along with Costas Christophi, Maria Deloria-Knoll, and myself. I also thank Doug Altman, Don Corle, Simon Day, Steven Hirschfeld, Damian McEntegart, Thomas Permutt, and Drummond Rennie for useful discussions of selection bias. I thank my former superiors Stephen Wilson, George Chi, Satya Dubey, Chuck Anello, Bob O'Neill, Tony Lachenbruch, and Susan Ellenberg, at the Food and Drug Administration, and more recently Phil Prorok and Peter Greenwald at the National Cancer Institute, for encouraging me to continue my research. I thank Jeffrey Mann for informing me of the randomization irregularities in the NINDS Stroke Trial, and

xii   *Preface*

Giuseppe Biondi-Zoccai for referring me to the information regarding the randomized trial of laparoscopic versus open appendectomy (Section 3.3.28). Finally, thanks also go to Rob Calver, Kathryn Sharples, and Jane Shepherd of John Wiley & Sons for helping the process to go smoothly.

Vance W. Berger
North Potomac, MD
October 2004

Part I
Is There a Problem with Reliability in Medical Studies?

1

An Evolution of Comparative Methodology

Our starting point is the desire to evaluate a new medical intervention to determine if it might be useful in clinical practice. We mean this to be quite inclusive, so the medical intervention might be a drug, a vaccine, a diet, a screening test, acupuncture, cognitive therapy, or anything else that might be used to promote health and/or treat or prevent a disease. Over the years, many different methods have been employed in the name of evaluating a new medical intervention (or evaluating a new use for an existing medical intervention). These methods include the design, conduct, and analysis of studies. In this chapter we critically evaluate some of the designs that have been used to evaluate medical interventions, filling in many of the details behind the outline provided by Smith (2003). We pay special attention to the existence and nature of the control group, and the method for determining who gets which treatments.

1.1 SINGLE-SUBJECT STUDIES

One might imagine a time when the standard evaluation of a medical intervention consisted in applying it to a single subject, and noting if this subject appeared to improve or deteriorate. Today we are well aware of the limitations of such single-subject studies, but it still may be useful to clarify what precisely these limitations are, to better understand the need for other methods. Consider a situation in which the natural history of a disease is known with absolute certainty, and

Selection Bias and Covariate Imbalances in Randomized Clinical Trials   V. W. Berger
© 2005 John Wiley & Sons, Ltd.

there is literally no variation across patients. In such a case, if one patient were demonstrated to deviate from this known path following treatment with a given agent, then the response would have to be attributed to the agent, and the study would be convincing. For example, if some new potion could be applied to a corpse and bring the corpse back to life, then one corpse would be sufficient to provide strong and convincing evidence.

It is common to refer to randomization as the basis for inference; see, for example, Berger (2000), Berger *et al.* (2002), and Berger and Bears (2003). In most cases it is true that randomization serves as the basis for inference. However, in the case just described, there is an alternative basis for inference. This is because we know the potential outcomes both with and without the potion. The former was observed and the latter follows from the certainty of the natural history in the absence of the potion. This makes the prediction causal, as opposed to probabilistic (Runde, 1996). With knowledge of both potential outcomes, we are in a position to perform causal inference, and compute a legitimate and valid p-value. The probability of observing so extreme a result under the null hypothesis that the potion does not work is exactly zero, so this is the p-value. Likewise, if at some point in the future we find ourselves in a position to conduct trials on clones, so that the assumption of exchangeability or identical potential outcomes (Greenland and Robins, 1986) becomes tenable, then randomization would be unnecessary as we would still be able to perform causal inference, and compute a legitimate and valid p-value of zero if any between-group difference is found.

1.2 CASE SERIES AND COHORT STUDIES

The conditions described above are not likely to be met in actual clinical practice. How well, then, does a single-subject study perform when the natural history is not known with certainty? The answer is not very well, because the experiences of a single subject represent the outcome of a single Bernoulli trial with unknown success probability. There is little hope, based on this single Bernoulli trial, of estimating the success probability, let alone of establishing its difference from the success probability in the absence of treatment. Intuitively, it is clear that a larger sample size will offer some benefits, because the sampling variability is reduced with increases in the sample size. As such, a case series or cohort of consecutive patients with the same or a similar diagnosis all treated the same way might be preferred to a single-subject study. These designs are often used for Phase II studies, to ascertain preliminary indications of efficacy.

While the case-series approach is certainly preferable (at least from a scientific perspective, but from an ethical perspective this point could be debated) to a single subject study, it also retains some of the same drawbacks of the single-subject study. In particular, the case-series design does not address the fact that the evaluation of a medical intervention is necessarily comparative. An individual may not care, for example, if a vaccine is 'good' or 'bad' in a vacuum, but this individual may have come to understand that these descriptors refer to the vaccine being better or worse, respectively, than the absence of the vaccine. Such an evaluation can be made only with a comparative study.

1.3 HISTORICAL CONTROLS

One of the simplest comparative designs is the historical control design, in which the experiences of a current case series are compared to the historical experiences of a prior cohort. This design allows for an assessment of 'better' or 'worse', which is certainly a strength. However, the comparison is confounded with both time trends and selection processes. That is, unintentionally or otherwise, especially good responders may be selected for the current case series. Clearly, this would bias the results in favor of the intervention used for the current case series. Conversely, especially bad responders may be selected for the current case series; this would bias the results in favor of the intervention used for the historical controls. Even if the current cohort is quite similar to the historical cohort prior to either being treated, they still may differ once treated for reasons having nothing to do with the treatments being compared. For example, ancillary care may be better now than it was in the past; this would bias the results in favor of the intervention used for the current case series. Conversely, managed care may deny the current cohort some health benefits that were available to the historical cohort; this would bias the results in favor of the intervention used for the historical controls. As we see, time itself is an important covariate that should be balanced across groups. This can be accomplished with parallel control groups, or groups that are treated at the same time.

1.4   PARALLEL CONTROL GROUPS

Many studies fall in the category of parallel control. For example, one could assess the effects of smoking by comparing those who do smoke today to those who do not. Each group might be followed up for some period of time, and any occurrences of cancer or heart disease would be noted, and compared across groups. This design would balance the effects of time across groups, and would therefore eliminate this source of bias. However, there are other sources of bias that remain, most notably self-selection bias. Consider that those subjects who choose to smoke may differ in important ways from those who do not. For example, they may engage in riskier behavior, or may drink more alcohol, or may eat fewer fruits and vegetables. This means that even if a clear difference between the experiences of the smokers and the experiences of the non-smokers is found, this difference may not be attributable to smoking itself. To attribute the differences in outcomes to the agents studied requires that the comparison groups be as comparable as possible in every way other than the difference in their treatments.

1.5   MATCHED STUDIES

The next improvement in study design is matching, including case–control studies (Breslow and Day, 1980). In our development, this can include both prospective and retrospective designs. As an example of the latter, the Los Angeles Retirement Study of Endometrial Cancer (Mack *et al.*, 1976; Breslow and Day, 1980, Section 5.1) was a case–control study designed to study the effect of exogenous estrogens on the risk of endometrial cancer. There were 63 cases of endometrial cancer identified from 1971 to 1975 in a retirement community near Los Angeles. Each case was matched to four controls, all of whom were alive and living in the community at the time the case (of endometrial cancer) was diagnosed, were born within one year of the case to whom they were matched, had the same marital status, and had entered the community at approximately the same time as the case. In addition, controls were chosen from among women who had not had a hysterectomy prior to the time the case was diagnosed and who were therefore still at risk for the disease. One purpose of the study was to determine whether gall bladder disease was associated with endometrial cancer.

Here, the search would have been retrospective, to find out which cases and which controls had experienced gall bladder disease that would have occurred prior to the present time. It is also possible to study the effects of gender, height, genetic profiles, exposure to particular carcinogens, or exposure to particular viruses with a prospective variation of the same design. One would find cases, defined, for example, as those having gall bladder disease, and then match each of these cases to some number of controls. Now all cases and controls could be followed prospectively to determine if they develop endometrial cancer. The basis for inference in this design is the exchangeability of the cases and the controls. That is, it is hoped that the cases and controls are identical to each other in every way other than the 'caseness', or that which makes the cases become cases and that which makes the controls remain controls (not cases).

It is possible, at least in theory, to match on any number of prognostic variables, so any number of variables can be balanced across cases in controls within each matched set. If all potential confounding variables are known and measured then, again at least in theory, randomization may be considered unnecessary (Villar and Carroli, 1996). The problem is that there are often prognostic variables that are not measured. For example, subjective health perceived by a patient can predict clinical outcomes and even mortality, even after adjusting for other observed predictors (Fayers and Sprangers, 2002). In fact, as Madersbacher *et al.* (2004) pointed out, 'The comparison of new treatment modalities with so-called matched controls and particularly historic controlled series . . . confuse the results by introducing errors resulting from case selection bias, stage migration, differences in follow-up, and the evolution of supportive care. These confounding, recall, and detection biases are particularly problematic for the results of oncologic trials because the respective surgical or medical therapies can be associated with considerable treatment-related morbidity.'

It is entirely possible that, in a case–control study, the matching does not balance the unknown and/or unmeasured covariates, such as subjective health perceptions. The same criticism applies to deterministic designs mistakenly referred to as 'randomized', such as minimization. For this reason, it is considered ideal to randomize, at least when doing so is feasible and ethical. Clearly, it is not always feasible or ethical to randomize. For example, it is not possible to randomize subjects to different genders (surgical interventions to modify the gender may be possible, but this is not the same as being born to a given gender). It is certainly conceivable to randomize subjects to exposure to carcinogens, but this is not ethical. As such, there is still a place for matched designs that are not randomized. From this point on, however, we restrict attention to randomized trials.

We note that alternating designs, and other deterministic designs including those in which allocations are based on the social security number, are often called 'randomized' (Berger and Bears, 2003), yet these are poor substitutes for true randomization. Nature is not in the business of randomizing the order in which patients show up to clinics, so there is no sense in which alternating designs represent truly randomized designs. As with all non-randomized studies, the observed data represent the only outcome that could have occurred. This means that if randomization is the basis for inference, and there was no true randomization, then the only valid p-value would have to assume the uninteresting value of 1.00. Only if the non-randomized study is performed in clones, or in a patient population whose natural history is known with certainty, would another basis for inference be available, to allow for the valid calculation of a more interesting p-value. Of course, assumptions can also serve as the basis for inference, as when a population is assumed to follow the normal distribution and sampling is assumed to be random.

While these two 'assumptions' are more often better described as violations of known facts to the contrary, a less objectionable assumption was discussed recently by Gallin *et al.* (2003) with regard to a study to evaluate the efficacy of itraconazole. Specifically, the treatments (itraconazole and placebo) were alternated over time periods within each patient, and this continued until the occurrence of an event or the end of the study. With the assumption that the probabilities of events in the two groups did not depend on time or exposure to prior treatments, it is possible to compute valid p-values, even in the absence of randomization. Now the study by Gallin *et al.* (2003) actually did use randomization, but it appears that the assumption of time homogeneity, and not randomization itself, served as the basis for inference and the calculation of p-values.

1.6 RANDOMIZATION

Randomization is often said to balance all covariates, at least in distribution, across the treatment groups. For example, Beller, Gebski, and Keech (2002) state that 'Allocation of participants to specific treatment groups in a random fashion ensures that each group is, on average, as alike as possible to the other group(s). The process of randomization aims to ensure similar levels of all risk factors in each group; not only known, but also unknown, characteristics are rendered comparable, resulting in similar numbers or levels of outcomes in each group, except for either the play of chance or a real effect of the intervention(s).' While it is certainly true that randomization is used for the purpose of ensuring comparability between or among comparison groups, we will see in Chapter 2 that it is categorically not true that this goal is achieved. However, it is worth reviewing the logic behind this statement to see where it can break down.

One basic tenet of most forms of randomization is that there is no opportunity for the subject to select a treatment, and no opportunity for the investigator to assign a treatment based on subject characteristics. Exceptions exist, at least to some extent; for example, the consumer principle of randomization would allow subjects to select not the treatment *per se* but rather the probability with which they are to receive each treatment (Bird, 2001). In most cases, however, there is no consumer choice, and allocation probabilities are determined in advance. Often, but not always, these probabilities are the same for all treatment groups, to achieve balance in sample sizes across the treatment groups. For simplicity, unless otherwise noted, we will consider only two-arm randomized trials with equal allocation probabilities to the two groups (1:1 randomization). Our development will apply more broadly, however, allowing for more treatment groups and unequal randomization.

The idea of randomization is to overlay a sequence of units (subjects, or patients) onto a sequence of treatment conditions. If neither sequence can influence the other, then there should be no bias in the assignment of the treatments, and the comparison groups should be comparable. We note that the bias from self-selection designs can be viewed as the influence of the former sequence on the latter sequence. Specifically, the identity of the units (or, in this case, patients) in the first sequence, and their ability to select treatments, not only influences but also determines the sequence of treatment assignments (Berger and Christophi, 2003). This is why it is not valid to compare the group of patients who were treated with (by virtue of having selected) one treatment to the group of patients who were treated with (by virtue of having selected) another treatment. In fact, sometimes there are contraindications that allow some patients to use one treatment but not another. In fact, eligibility for a chemotherapy protocol was recently found to be a good prognostic factor for invasive bladder cancer after radical cystectomy (Madersbacher *et al.*, 2004).

With randomization (understood to exclude the consumer variety), there should be no such influence of the subjects on the treatment assignments. The veracity of this statement depends on the nature of the randomization procedure. Consider, for example, randomization by tossing coins. If the coin tossing takes place only after the subject to be randomized has been identified, then it would be possible to take into consideration the preferences of this subject by rejecting the outcome of the coin toss until the preferred outcome is observed.

Schulz (1995a) defines randomization as follows: 'First, an unpredictable allocation sequence must be generated based on a random procedure. Second, strict implementation of that schedule must be secured through an assignment mechanism (allocation concealment process) that prevents foreknowledge of the treatment assignment'. He goes on to call it a 'mistake' that many medical researchers regard only the sequence generation process as the randomization itself. We disagree, and find good reason to follow instead Berger and Bears (2003) in defining randomization strictly on how the allocation sequence is generated, randomly or not. That is, a trial is randomized if, and only if, the accession numbers from any one treatment group constitute a random sample from the set of all accession numbers used. In taking this as the definition of randomization, we are not denying the importance of allocation concealment, but, for reasons that will become clear in the remainder of this book and especially in Chapter 2, there are good reasons to regard the two processes, randomization and allocation concealment, as markedly distinct entities, to allow for consideration of one without the other.

1.7 ADVANCE RANDOMIZATION

The type of subversion considered in the previous paragraph would represent a breakdown in the integrity of the randomization itself (later we will discuss other subversions of the allocation process that have nothing to do with any breakdown in the randomization itself), and can occur in any trial for which randomization takes place only after the (human) subjects were already selected. In practice, not only are the allocation proportions determined prior to the initiation of patient recruitment, but in fact the allocation sequence itself is also determined in advance, prior to the initiation of patient recruitment. This design feature ensures that the sequence of subjects to be randomized cannot influence the sequence of treatment assignments (Berger and Christophi, 2003). In this sense, randomization has served its purpose. We can say that randomization has contributed to the balancing of both measured and unmeasured covariates. Certainly, it has done a better job of this than any deterministic design could (Moses, 1995). Is it possible, however, for the direction of the influence to be reversed? That is, can the sequence of treatment assignments influence the sequence of subjects to be randomized?

At first, this notion seems preposterous. How can a treatment assignment alter the baseline characteristics (pre-randomization) of a patient? However, we need to take a broader view of this potential influence than simply the influence of a given treatment assignment on the corresponding patient. In fact, it is clear that once the patient is selected to be randomized, there can be no influence of the treatment allocation on that patient (at least not on any patient characteristics prior to randomization). However, if it is known that the next allocation will be to a given treatment group, then this advance knowledge may lead to selective patient recruitment. This concern can be addressed, of course, by determining each allocation only after the patient to be enrolled is identified, as was suggested by Clarke (2002). But either the allocation to be made or the patient to be enrolled has to be selected first; whichever it is may influence the other.

The biases possible with randomization only after patient accrual are at least as serious as the biases possible with advance randomization, so the best approach seems to be to randomize first, and then recruit the patients, but to try to do so in such a way that the influence of the treatment assignments on the patients enrolled is minimized, or preferably eliminated. Berger and Christophi (2003) enumerated some conditions under which this reverse influence can be completely eliminated. For example, if the trial is performed in clones, all of whom are identical in every way to each other, then there can be no preferential patient selection to any treatment group. If all eligible patients must be enrolled, and can neither refuse consent (possibly after being discouraged by an investigator who was aware of the upcoming treatment assignment) or denied enrollment, then there would be no way for the treatment assignments to influence the patient selection (although it could, of course, influence the patient evaluation after enrollment).

In practice, however, both investigators and patients enjoy enrollment discretion, and studies are not done in clones. Still, if the patients to be randomized can be all assembled at once, prior to randomization, and then randomized all at once, then there is no opportunity to act on any advance knowledge. However, most trials are sequential, in the sense of using staggered patient entry, and patients are randomized as they are enrolled, often due to the need for immediate treatment of the disease that qualified them for the trial in the first place. This leaves one other hope for eliminating the influence of the treatment assignments on the patient selection. If there is absolutely no advance knowledge of upcoming allocations, then there is no opportunity to preferentially select better responders into one treatment group or the other. This is the idea behind allocation concealment (Schulz, 1995a, 1995b, 1996), which is essentially the masking (or concealing) of each allocation just until it is executed. That is, if the allocation itself reveals the nature of the treatment assigned, this would not constitute a violation of allocation concealment, because it occurs only after the patient to be allocated has already been selected.

1.8   ALLOCATION CONCEALMENT

Discussions of the imperfections of masking are quite relevant to the evaluation of the success of allocation concealment. For example, in a discussion of the distinction between a claim of masking and true masking, Oxtoby *et al.* (1989) pointed out that 'the presumption that a plan to which one has aspired has come to fruition by virtue of aspiration alone is not science, and is particularly inapposite for a profession which should have a reputation for making clear distinctions between fantasy and reality'. Masking may be defined as either the process (researchers not revealing treatment codes until the database is locked) or the result (complete ignorance of all trial participants as to which patients received which treatments until the database is locked). Analogously, then, allocation concealment may be defined as either the process (researchers not revealing treatment codes until the patient is randomized) or the result (complete ignorance of all trial participants as to which patients received which treatments until the patient is randomized). It is often said that masking is possible only some of the time, while allocation concealment is always possible. The reason for this sentiment is clear enough. It is hard to imagine how to mask a trial comparing a surgery to a medical treatment, for example. Yet allocation concealment would still be possible even in this case, because the unmasking of each patient would occur only after the allocation, and after the selection of the patient.

Of course, there are cases in which sham surgery is ethical, and might cause the study to be as well masked as trials comparing a medicine to a placebo (Jones *et al.*, 2003). Yet even in cases in which sham surgery is deemed unethical, there is still something troubling about stating that masking is possible only some of the time, while allocation concealment is always possible. Specifically, Berger and Christophi (2003) noted that the process of masking is always possible, and pointed out that

this confusion of the two definitions is a double-standard. If masking is possible only some of the time, then clearly reference is being made to the result, and not the process. To be fair, then, one would have to ask if the *result* of allocation concealment is always possible. Sealed envelopes have been held to lights, phantom patients have been enrolled, and locked files have been raided to determine upcoming treatment allocations in successful subversions of allocation concealment (Schulz, 1995a) ... so only the *process* of allocation concealment, but not its result, can be ensured.

In future chapters, we will have more to say about the specific mechanisms by which allocation concealment can be subverted. For now, we highlight the two key points, which are as follows. First, even in trials labeled as 'randomized', randomization can be conducted with or without error, or not at all. Second, even in trials that are properly randomized, without error or subversion, a lack of the result of

allocation concealment can occur even when the trial claims allocation concealment (the process). Hence, baseline covariate imbalances
across treatment groups can by systematic even in such trials. We
call such systematic baseline covariate imbalances across treatment
groups selection bias, although the term 'selection bias' has come to
have many different meanings in different contexts (Mark, 1997).

Our interest in selection bias is confined to the type of selection bias
that interferes with internal validity, or a fair and unbiased comparison of the treatment groups. The mechanism for this type of selection
bias is most easily understood in the context of non-randomized designs, and especially self-selection designs. It is commonly believed
that randomization by itself will eliminate this type of selection bias,
but in fact, as we have seen and will explore further in later chapters, it does not. Moreover, such selection bias can occur even when
the randomization was performed successfully, and not subverted. Of
course, the randomization itself may be subverted too. For example,
as we have seen, randomization may occur only after patient selection, and this can degenerate into what essentially becomes a *de facto*
non-randomized trial.

1.9   RESIDUAL SELECTION BIAS

We refer to the selection bias that interferes with internal validity in
randomized trials with patient selection preceding randomization as
first-order residual selection bias, to distinguish it from its related form
that occurs in non-randomized studies. It may be tempting to believe
that simply performing the randomization in advance would eliminate all such selection bias, but this is not true either, as future allocations may be predictable. We refer to the selection bias that interferes
with internal validity in *advance* randomized trials as second-order
residual selection bias. It may be tempting to believe that allocation
concealment would eliminate all such selection bias. Indeed, the objective of allocation concealment would eliminate all such selection
bias, but again, the claim of allocation concealment refers to the process, and this is not sufficient to ensure that allocation concealment
has achieved its objective. We refer to the selection bias that interferes with internal validity in advance randomized trials with imperfect or unsuccessful (subverted) allocation concealment as third-order

residual selection bias. This will be the type of selection bias with which we are most concerned, as it is the only one that does not have a simple countermeasure that is generally known and utilized.

In Chapter 2 we will discuss, in greater detail, the mechanisms by which selection bias may occur even in properly randomized trials, at least as we define the term 'properly randomized'. In Chapter 3 we will provide some evidence that this type of selection bias actually occurs, and is not merely a hypothetical concern. In Chapter 4 we will discuss the impact one can expect selection bias to have on the results of trials. In Chapter 5 we will discuss measures (beyond randomization) that can be taken to prevent selection bias. In Chapter 6 we will discuss methods by which selection bias can be detected, or hopefully ruled out, from any given randomized trial. In Chapter 7 we will discuss methods that can be used to adjust for selection bias, in case it is found but useful between-group comparisons need to be salvaged anyway. In Chapter 8 we will summarize the overall recommendations for managing selection bias in randomized trials.

2

Susceptibility of Randomized Trials to Subversion and Selection Bias

It is widely believed that baseline imbalances in randomized clinical trials must necessarily be random. In fact, there is a vast literature (see, for example, Senn, 1994) which indicates that because baseline imbalances in randomized clinical trials must necessarily be random, hypothesis testing for baseline imbalance is illogical in randomized trials. The argument is that any hypothesis test would be testing for balance in the population of all possible randomizations which could have occurred. By virtue of the random allocation of subjects to treatment groups, this is necessarily true. Thus, any time this null hypothesis is rejected we know that we have a Type I error. This argument would of course be compelling if all observed baseline imbalances in the context of randomized trials were necessarily random. But is this actually the case? That is, are there mechanisms by which patients with specific covariates may be selected for inclusion into a particular treatment group even among randomized trials? Schulz (1995a) and Torgerson and Roberts (1999) argue that in fact systematic baseline imbalances (selection bias) can occur in randomized trials. If this is true, then this selection bias would force imbalance in those covariates, measured or unmeasured, that are used for the patient selection.

Selection Bias and Covariate Imbalances in Randomized Clinical Trials   V. W. Berger
© 2005 John Wiley & Sons, Ltd.

Clearly, the imbalance would not be random, because it would occur again if the trial were repeated under the same conditions. We will find that selection bias is possible not only in randomized trials, but in fact also in masked randomized trials conducted with allocation concealment. This being the case, the view that one ought not to test formally for baseline imbalance in randomized trials confuses the sufficiency of randomization to eliminate systematic imbalances with necessity. That is, randomization may be necessary to ensure that any observed baseline imbalances are random, but it certainly is not sufficient, as selection bias can occur even in randomized trials. While not generally cast specifically as a test of this type of selection bias in randomized trials, tests of baseline imbalances do, in fact, constitute tests for selection bias, although they are crude ones.

2.1 CAN RANDOMIZED TRIALS BE SUBVERTED?

The term 'selection bias' has been used to describe many different biases (Mark, 1997). To focus ideas, we confine our attention to the types of selection bias that interfere with internal validity (a fair comparison); that is, we do not consider external validity. Groups of patients to be compared may differ in important ways even before any intervention is applied (Prorok *et al.*, 1981). These baseline imbalances cannot be attributed to the interventions, but they can interfere with and overwhelm the comparison of the interventions (Green and Byar, 1984). If treatments are independent of patient characteristics, then any baseline imbalances (even if statistically significant) are due to chance variation only. This is one reason often cited for using randomization. On the other hand, a systematic explanation for the imbalances, known or unknown, would constitute selection bias, even if the imbalances are not statistically significant, or even readily observed (Berger and Exner, 1999). Berger and Christophi (2003) present a sequence of mechanisms by which selection bias may occur, starting with observational studies, and such countermeasures as randomization, allocation concealment, and masking (see Table 2.1). While we focus only on randomized trials here, the consideration of the mechanism for selection bias in observational studies can still be instructive.

Table 2.1 Selection Bias with Randomization and Allocation Concealment reproduced from Journal of Modern Applied Statistical Methods, 2003, Vol. 2, No. 1, p. 85

		{(A C C A); (C C A A)}		{(A C A C); (C A A C)}	
S	P{active} Range*	P{active}	Randomized	P{active}	Randomized
S1	[0.50,1.00]	0.50	Active	0.50	Active
S2	[0.00,0.33]	0.33	Control	0.33	Control
S3	[1.00,1.00]	0.50	–	0.50	–
S4	[0.00,0.50]	0.50	Control	0.50	Active
S5	[0.50,1.00]	1.00	Active	0.00	–
S6	[0.00,0.50]	0.50	Control	0.00	Control
S7	[0.00,0.50]	0.67	–	0.50	Control
S8	[0.67,1.00]	0.67	Control	0.67	Active
S9	[0.67,1.00]	1.00	Active	0.50	–
S10	[0.00,0.50]	1.00	–	0.50	Active
S11	[0.33,0.67]	1.00	–	0.00	–
S12	[0.00,1.00]	1.00	Active	0.00	Control

*The range of P{active} values for which the patient gets randomized. P{active} computed according to the formula of Berger and Exner [3] using the randomized block procedure with a fixed block size of four. Not only does treatment assignment for randomized patients depend upon the allocation sequence, but in fact Patients #S5, #S7, #S9, and #S10 may or may not be randomized depending on the allocation sequence, and Patient #S3 cannot get the control.

In observational studies, investigators may (and, in clinical practice, do) assign treatments based on patient characteristics (Green and Byar, 1984; Rubin, 1977). Among over-the-counter treatments, patients may (and generally do) select their own treatment. With consumer randomization (Bird, 2001), patients select not their treatment *per se* but rather their randomization probability, at least from among a given set of choices. We see that allocation discretion, or the ability to select a treatment, may be available to the patient, to the investigator, or to both the patient and the investigator. For example, if an antibiotic is available only with a prescription, and a physician is reluctant to prescribe it based on the belief that the symptoms reflect a viral infection, but the patient insists, then a negotiation would likely ensue, and both parties would exercise some degree of allocation discretion.

Clearly, those patients selecting one treatment (or probability of receiving a given treatment) may differ systematically from those selecting another (Green and Byar, 1984). One countermeasure to prevent patient characteristics from influencing the allocation sequence, through either overt treatment assignment based on patient characteristics or self-selection, is dictated allocation, which essentially is the lack of freedom of choice. That is, a treatment is assigned, randomly or otherwise, independent of any patient characteristics (including specific treatment preferences). This dictated allocation is not inherent in randomized trials, but in practice it is one of the cornerstones of randomization as implemented. If allocation is alternated, for example, then either all patients with even accession numbers or all patients with odd accession numbers receive the active or experimental treatment (E). The others receive the control (C). This non-randomized allocation procedure (which is often incorrectly referred to as randomized; see Berger and Bears, 2003) is one form of dictated allocation, and so it would prevent the type of selection bias considered above. But would it prevent all selection bias?

In fact, this step does not even eliminate all self-selection bias, because many randomized trials are conducted in various medical centers (clinics). When this is the case, it is typical to stratify the allocation by center, meaning that each center has its own independent allocation sequence. Sometimes, there is more than one center in a given metropolitan area. When this is the case, and the trial is unmasked, it is possible for a patient to receive the control treatment (or, more generally, any treatment other than the one he or she prefers), drop out of the trial, and then enter the trial again at a different center, in hopes of receiving the preferred treatment. In fact, this has been known to occur (Brauer, 2004).

Another threat to the sanctity of randomization arises because of mistakes that may occur. See, for example, the COMET Study Group (2001). A third potential problem involves limited quantities of drug at the centers that enroll patients. McEntegart (2003) points out that when supplies of one treatment, but not all treatments, are out of stock there are three options that can be used. First, the randomization can be halted. That is, recruitment can be stopped until there are adequate supplies of all treatments. Second, recruitment can proceed until a patient is randomized to the treatment which needs to be ordered, at which point it stops. Third, randomization can continue, and when a patient is randomized to a treatment that needs to be ordered, a switch is made to the next accession number that corresponds to a treatment that is in stock (which is called 'forcing').

Depending on how this situation is handled, there could be the potential for unmasking and selection bias. While the selection bias arising due to either limited stock of some treatments or patients enrolling at more than one center in hopes of obtaining their preferred treatment is a fairly serious problem, we focus instead on selection bias in randomized trials caused by investigator (not sponsor or patient) actions.

Typically, accrual in prospective medical studies is sequential, meaning that the patients do not enter the study at the same time. There is also generally enrollment discretion (Chalmers, 1990), which allows an investigator to deny enrollment to certain patients at his or her discretion. Penston (2003, page 68) listed the exclusion criteria of five major trials, and emphasized how imprecise they were. These exclusion criteria included 'serious organic or psychiatric disease' for the ASSET and LATE Trials, 'some other life-threatening disease' for the ISIS-2 Trial, 'not specified by the protocol but by the responsible physician' for the ISIS-3 Trial, and 'any other disorder that the investigator judged would place the patient at increased risk' for the ASSENT-2 Trial. Clearly, then, discretion exists, as it is not the case that every potential allocation decision is to be made according to some objective and explicit criterion in the protocol.

The combination of sequential accrual and enrollment discretion lays the foundation for selection bias, but of a different type from that considered previously. If there is advance knowledge of the upcoming treatment, then it would be possible for an investigator to deny enrollment to patients lacking the characteristics that would make them 'suitable' to receive this known upcoming treatment (Schulz, 1995a; Schulz and Grimes, 2002a). The sequential accrual contributes to the likelihood of advance knowledge of the allocation sequence, as we will see. Because two of the conditions (sequential accrual and enrollment discretion) that enable this type of selection bias cannot be modified in most trials, it is the third condition, the predictable allocation sequence (Schulz and Grimes, 2002b), that needs to be kept in check to prevent this form of selection bias. Does randomization accomplish this objective, either by itself or in conjunction with masking and/or allocation concealment?

Recall that randomization is not one specific method by which allocation takes place. Rather, it refers to a collection of many different methods. For our purposes, an allocation procedure is randomized if each accession number has the same probability of being matched to any given treatment. Three points are worth mentioning. First, this definition is overly restrictive, in that designs in which the allocation proportions vary over the course of the trial may still be randomized, but we will not consider these designs here. Second, what is common is the probability of a given treatment allocation as the accession numbers vary. There is no requirement that for any given accession number each treatment be equally likely. Third, the common probabilities are unconditional. Once we condition on what has occurred so far in the trial, there is no requirement that any of these probabilities be common.

One form of randomization is urn randomization (Wei and Lachin, 1988), which is conducted by tossing a (possibly biased) coin each time a patient is to be allocated. Heads indicates that the experimental treatment is to be received, and tails indicates that the control treatment is to be received, or vice versa. With urn randomization, there is no actual allocation discretion, yet having screened and evaluated a given patient, the investigator might exercise *de facto* allocation discretion to reject the toss and repeat until the preferred allocation is observed. Clearly, this would constitute both a subversion of the randomization and selection bias. So the answer to the question that defines this section is an emphatic yes. Randomization by itself can be subverted.

2.2 IF RANDOMIZED TRIALS ARE SUBVERTED, DO THEY CEASE TO BE RANDOMIZED TRIALS?

In the previous section we saw that selection bias may occur even in a randomized trial if the randomization is based on urns. But the mechanism for subverting the randomization actually constitutes a negation of this randomization process itself. That is, the randomization is what is broken, in that it is no longer the case that each accession number has the same probability of being matched to a given treatment. As such, one may prefer not to call this a randomized trial. More generally, it may seem that if any subversion occurs in any randomized trial, based on urns or not, then the trial was not truly randomized. This statement implies that randomization confers absolute protection against any subversion, so that any covariate imbalances must be random. But now consider another mechanism for selection bias in a randomized trial with dictated allocation.

With minimization, or dynamic randomization (Pocock and Simon, 1975), the allocation is determined by minimizing an imbalance function, and randomization may be used to break the ties. So there is both dictated allocation and randomization. Yet because most allocations will be deterministic, it would be possible to determine the allocation to be made once a patient has been identified. A patient enrollment decision may then be based on a combination of the treatment to be assigned and values of observed covariates that were not used to define the imbalance function. Clearly, selection bias is possible here, too, yet this would remain a randomized trial.

We do not concern ourselves further with minimization, but note that even with more common methods of randomization, such as randomized blocks, selection bias is possible without the randomization itself being the root of the problem. The distinction is not mere semantics. To develop this distinction, we follow Berger and Christophi (2003) and consider randomization to be *conventional* if the allocation sequence is generated in advance of screening any patients, and *unconventional* otherwise. Clearly, conventional randomization prevents many types of selection bias. But again, selection bias may result from enrollment discretion and advance knowledge of the allocation sequence; the latter may be facilitated by conventional randomization, as the allocation sequence may be posted publicly before patients are screened (Schulz and Grimes, 2002a). In such a case, the randomization itself has not failed, because it correctly (in the absence of any other subversions) assigns accession numbers (hence enrolled patients) to treatment groups based on the prospective allocation sequence according to the definition in Chapter 1.

That is, prior to identifying the patient to assume the role of patient 1, and the patient to assume the role of patient 2, and so on, patient 1 has a given probability of receiving the experimental treatment. This probability is known, and is determined by the randomization procedure. Patient 2, and every other patient, has the same probability of receiving the experimental treatment. But these probabilities are unconditional, and apply only up to the point when the patients who will assume the accession numbers become identified. If there is no systematic association between accession numbers and patient characteristics, then by randomization the accession numbers in any one treatment group would constitute a random sample of all the accession numbers that were used, and so any function of the accession numbers will be truly equally distributed across the treatment groups. For example, the mean accession number within treatment groups will differ across treatment groups only by chance alone. This sounds very much like the argument proposed at the beginning of this chapter, regarding *any* baseline characteristic being truly equally distributed across the treatment groups. Indeed, were patients matched or linked to accession numbers *prior* to the accession numbers being randomly allocated to the treatment groups, then this would be a valid argument. But this extension breaks down upon scrutiny in sequential trials.

The problem, as illustrated in Table 2.1 of Berger and Christophi (2003), is in the matching of eligible patients to accession numbers, which can be accomplished through strategic selection or rejection of patients based on a combination of the upcoming treatment to be assigned and specific patient characteristics. One could imagine, for example, randomly picking two of the first four accession numbers to be called group A, and the other two being called group B. But if one then forces women to occupy the randomly chosen accession numbers corresponding to group A, and forces men to occupy the randomly chosen accession numbers corresponding to group B, then the gender compositions will certainly differ across groups, and this difference will certainly not be random. While the accession numbers of group A would truly be a random sample of all used accession numbers, the genders of group A would not be a random sample of the genders. Herein lies the basis for baseline testing in randomized trials. While randomized trials, if properly randomized, guarantee that the accession numbers in any one treatment arm constitute a random sample of all the used accession numbers, this does not imply that the distribution of any given patient characteristic within one treatment group is a random sample of that same patient characteristic over all patients randomized. Moreover, it is this latter hypothesis, concerning the proper randomization of a given patient characteristic, and not the former one concerning accession numbers, that forms the null hypothesis for a baseline test of imbalance.

A countermeasure to eliminate the advance knowledge of upcoming allocations that leads to this type of selection bias is that each allocation be determined only after the patient to be enrolled is identified (Clarke, 2002), as occurs with minimization (Pocock and Simon,

1975). The fundamental problem, however, is that either the allocation to be made or the patient to be enrolled has to be selected first;
whichever it is may influence the other, and the biases possible with
unconventional randomization are at least as serious as the biases
possible with conventional randomization. The bottom line is that selection bias may occur even in properly randomized trials. We will
discuss this type of selection bias further, in relation to masking and
allocation concealment, in the next sections.

2.3 WHAT IS MASKING?

The mechanism we provided for selection bias in a properly randomized trial is a little removed from actual practice, as it is a clear violation
of masking. If such a fictitious example were needed to make the point,
then there would not be much to talk about. However, further insight
into masking and allocation concealment (to be discussed in the next
section) reveals that the problem may well persist even when these
measures are used. In a discussion of the distinction between a claim
of masking and true masking, Oxtoby *et al.* (1989) pointed out that
'the presumption that a plan to which one has aspired has come to
fruition by virtue of aspiration alone is not science, and is particularly
inapposite for a profession which should have a reputation for making
clear distinctions between fantasy and reality'. This profound remark
highlights the distinction between an action and its effect. Bearing in
mind the fact that the effect of an action may differ from its objective,
it would not be technically correct to state that the dishes are clean
just because they were washed. It would be more correct to state that
an action was taken specifically to ensure that the dishes are clean,
but without checking the success of the action, we would remain in
doubt as to whether or not the objective was attained. Likewise, one
can vaccinate a child against influenza, but this is distinct from stating
that this child will not contract influenza. The same considerations
apply to masking (Fergusson *et al.*, 2004a).

Masking may be defined as either a process (researchers not revealing treatment codes until the database is locked) or as a result
(complete ignorance of all trial participants as to which patients received which treatments). The objective of masking is the result of
complete ignorance, but a masking claim can indicate only that the

process was used. The use of this masking process may help to ensure the ignorance of some parties with regard to treatment allocations, but is unlikely to ensure the desired state of complete ignorance of all parties. As the legal term 'inevitable discovery' suggests, knowledge transfers by various mechanisms (Berger and Christophi, 2003). There are many mechanisms by which a trial planned as masked, and referred to in subsequent reports as masked, may become unmasked, either partially or completely (Carroll *et al.*, 1994). This includes intentional unmasking for safety reasons, distinguishing features of the treatments, and the detection of similarity to or difference from a treatment known to have been administered during a single-masked pre-randomization run-in period.

Mitchell (1981) pointed out that 'a truly double-blind study cannot be carried out with beta-blockers. Doctors only have to count a patient's pulse, measure his blood pressure, and ask him whether his hands are cold to make a very shrewd guess about his treatment group.' More recently, Martin *et al.* (2003) point out that 'Depending on the way in which the sham is delivered, the physical sensation experienced can differ when receiving sham and active treatment, effectively unblinding the patient'.

Penston (2003, page 40) adds 'blinding may simply be impossible: this is most obviously the case with surgical procedures but also applies to many other types of treatment including, for instance, orthopaedic traction, physiotherapy, and psychotherapy. And, even when it is suitable, the double-blind technique may fail to conceal the treatment from patients or investigators. Physiological effects of drugs may inform the investigators of the treatment, for example, bradycardia with beta-blockers or tremor with salbutamol. If a drug has well-recognised and obvious adverse reactions – for example, the extra-pyramidal side-effects of phenothiazines or the facial changes associated with prednisolone – then the occurrence of these problems discloses the presence of the drug. There are also incidental features of certain drugs that signal their use – the orange colour of the urine of patients receiving rifampicin and the black discolouration of the stools in those taking iron or bismuth compounds being obvious examples. Finally, the double-blind process is susceptible to fraud: details of the actual treatment received by individual patients may be obtained by tampering with envelopes or breaking the randomization codes, as well as testing tablets obtained while assessing compliance during follow-up visits.'

A cynical view, but one that is hard to refute, is that while an un-masked trial is what it is understood to be – each allocation is known once it is made – a masked trial differs from an unmasked trial only quantitatively. That is, a masked trial can be characterized by the number of unmasked allocations, and the timing of the unmasking of each of these allocations, being unknown. Unfortunately, the common usage of the term suggests, to the contrary, that there is no unmasking at all in trials labeled as 'masked'.

2.4 WHAT IS ALLOCATION CONCEALMENT?

Allocation concealment was described in Chapter 1, but there is still a benefit to discussing it in greater detail. We do so now. Unconventional randomization may not be able to eliminate advance knowledge of patient characteristics, but one might hope to eliminate advance knowledge of the allocation sequence with a combination of conventional randomization and allocation concealment, which is often taken to mean precisely this lack of advance knowledge. Technically, allocation concealment (Schulz, 1995a, 1995b, 1996) is essentially the masking of each allocation just until it is executed. At first glance, allocation concealment seems to be quite similar to masking, and the two may be confused in practice. The key distinction is the timing of the unmasking. With masking, the unmasking should occur subsequent to the finalization of the database, after all measurements have been recorded and checked. Allocation concealment is less restrictive, as it requires masking only until the patient to be allocated is selected, and then allocated to the appropriate treatment group (that is, only until the patient is matched to the accession number); certainly there is no requirement that the masking extend into the post-randomization measurement phase. Allocation concealment appears, on the surface, to be the answer to our prayers. Not only would it eliminate the possibility of the type of selection bias we described by eliminating advance knowledge of upcoming allocations, but it also appears to be a step that one can take in practice. That is, masking is often said to be possible only some of the time, while allocation concealment is always possible (Schulz, 1995a, 1995b, 1996). We will examine this statement carefully in the next section, and conclude that it is false.

2.5   A DOUBLE STANDARD

Consider more carefully the statement that masking is possible only some of the time, while allocation concealment is always possible. For example, Schulz (1995a) states that: 'Allocation concealment should not be confused with blinding. Allocation concealment seeks to prevent selection bias, protects the assignment sequence before and until allocation, and can always be successfully implemented. In contrast, blinding seeks to prevent ascertainment bias, protects the sequence after allocation, and cannot always be implemented'. The classical example of a trial that cannot be masked is a surgical trial, although care is required in this regard. McCulloch *et al.* (2002) distinguished three types of surgical trial, with Type I comparing medical treatments in surgical patients. These are not the trials that are generally claimed to be impossible to mask. Type 2 trials compare surgical techniques, and Type 3 trials compare a surgical technique to a medical or other non-surgical technique. These are the trials at the heart of the claim that masking is possible only some of the time. But is this claim legitimate?

If masking is possible only some of the time, then clearly reference is being made to the result of masking, and not the process of masking. After all, it is always possible to mask the trial if by this we mean only that the researchers do not intentionally reveal the allocations to the investigators or the patients. That is, the process of masking is always possible. And it is true that the process of allocation concealment is also always possible. But for a fair comparison, one would have to ask if the *result* of allocation concealment is always possible. Sealed envelopes have been held to lights, phantom patients have been enrolled, and locked files have been raided to determine upcoming treatment allocations in successful subversions of allocation concealment (Schulz, 1995a). Also, it may be clear what a given patient would receive, if enrolled, if cluster randomization (Jordhoy *et al.*, 2002) or minimization (Pocock and Simon, 1975) is used. Drug bottle numbers can also lead to prediction (Kuznetsova, 2002). It is not possible to enumerate, and rule out, all the mechanisms by which allocations can be observed. We are not prepared to take the success of allocation concealment on faith in an actual trial, so only the *process* of allocation concealment, but not its result, can be ensured.

We see that the process of masking is always possible, the process of allocation concealment is always possible, the results of masking are not always possible, and the results of allocation concealment are not always possible. It would seem, then, to be a double standard to take the more rigid definition of masking and the less rigid definition of allocation concealment and conclude that allocation concealment is always possible while masking is not. Yet we will intentionally step right into this trap, and proceed as if masking cannot be completely reliable in practice, but allocation concealment is. We do so for the purpose of this chapter to demonstrate that even in this unrealistically optimistic case, subversion is still possible.

2.6 WHAT IF ALLOCATION CONCEALMENT COULD BE ENSURED?

We saw, in the previous section, that the process of masking is always possible, the process of allocation concealment is always possible, the results of masking are not always possible, and the results of allocation concealment are not always possible. Yet from this point on we assume, quite unrealistically, that no allocations can be observed prior to being executed. Does this mean that they cannot be predicted, perfectly or imperfectly? Consider that allocation proportions may be changed during the conduct of the study. Sometimes this occurs inadvertently (see Lippman *et al.*, 2001), but more often the change will be planned. For example, in a randomized depression study of nurse telehealth care (Hunkeler *et al.*, 2000), the initial 40:60 randomization to two groups later became 40:20 to those same two groups, with the remaining 40% allocated to a new third group. Knowing that more late patients than early patients would be allocated to the third group constitutes advance knowledge of the allocations which, though imperfect, allows for deferred enrollment (Schulz, 1996) of those subjects most 'suitable' for the third group until after the new proportions took effect.

In the majority of randomized trials, the allocation proportions remain fixed throughout the duration of the trial. Would this, in conjunction with allocation concealment, eliminate selection bias? Randomization is *unrestricted* (Schulz and Grimes, 2002b) if a patient's likelihood of receiving either treatment is independent of all previous allocations, otherwise it is *restricted* (ter Riet and Kessels, 1995). For example, the random allocation rule (Schulz and Grimes, 2002b) requires that both treatment groups be assigned equally often. This is one form of restricted randomization, as the final allocation would be determined by the prior ones. Even with allocation concealment and fixed allocation proportions, patterns created by restrictions on the randomization allow prediction of the allocation sequence. Berger and Exner (1999) quantified this extent of advance knowledge with the conditional probability, P{E}, of a given patient being allocated to the experimental group given the previous allocations. With 1:1 allocation, P{E} = 0.5 for the first patient; with alternation, P{E} is always either 0 or 1. Note that P{E} reflects the restrictions on the allocation sequences, and becomes a patient characteristic only after that patient is randomized.

With enrollment discretion, P{E} may be used, in conjunction with the estimated potential outcomes of each patient to each treatment, say $\mathbf{Y} = \{Y(E), Y(C)\}$ for the active and control treatments, respectively, as a basis for enrollment decisions. Any baseline characteristic, including gender, age, race, and pre-existing medical conditions, may be considered in deriving the value of $\mathbf{Y}$ for a given patient. Based on $\mathbf{Y}$, the investigator might select a range of P{E} values for which the patient would be enrolled. If the P{E} value at the time this patient is screened happens to fall outside of this patient's P{E} range, then the patient will be denied enrollment, and another patient will be screened. Only when a patient is found with a P{E} range to match the actual P{E} value will the patient be enrolled. Selection bias occurs if the P{E} range is restricted based on $\mathbf{Y}$. It would be possible, for example, to enroll patients only if P{E} and $\mathbf{Y}$ are both large (suppose that larger $\mathbf{Y}$ values indicate better responses) or both small, but not if they are discordant (Schulz, 1995a). This possibility is depicted in Table 2.2 of Berger and Christophi (2003), using randomized blocks of size 4 to calculate P{E} (Berger and Exner, 1999). Figure 2.1 also demonstrates how this mechanism for selection bias works, in the case of blocks of size 2. Panel A shows the flow of patients where there is neither selection bias nor any random imbalance. This is the ideal situation in which there is perfect balance, both in each P{E} group and in each treatment group. Panel B shows a random imbalance, in which by chance more females end up in the P{E} = 0.0 group and more males end up in the P{E} = 1.0 group. This imbalance is

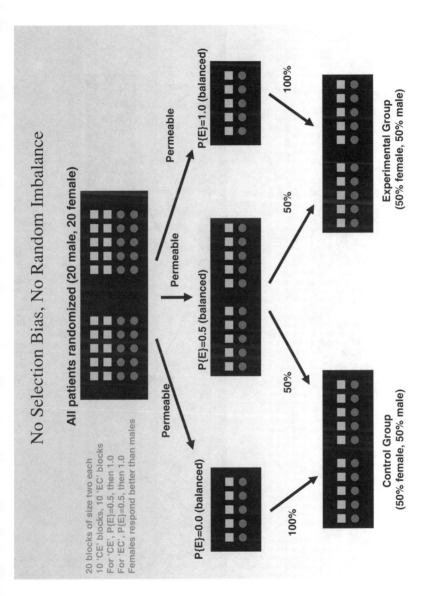

Figure 2.1

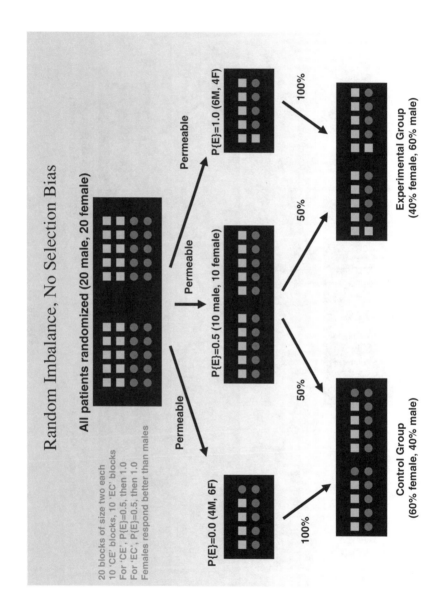

Figure 2.1 (*Continued*)

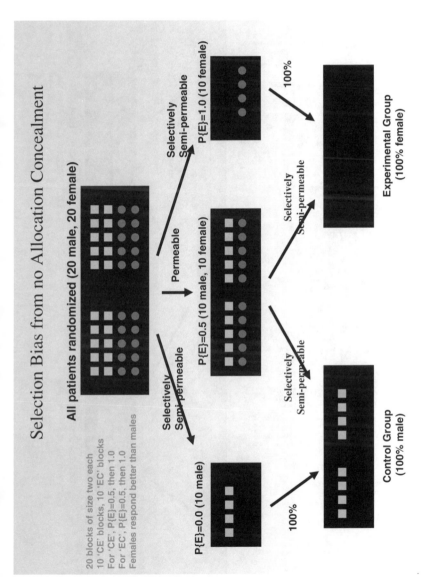

Figure 2.1 *(Continued)*

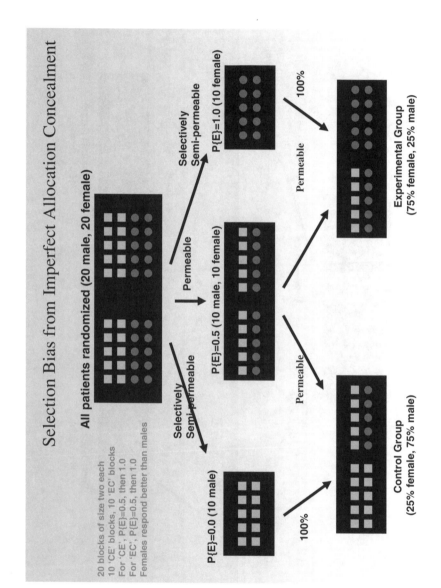

Figure 2.1 (*Continued*)

retained in the ultimate treatment groups, so the active group has more males than the control group does. Panel C shows a complete lack of any allocation concealment, so that the upcoming allocations need not be predicted – they can be directly observed no matter what the value of P{E}. In this rather extreme case, the entire control group is male and the entire active group is female. Finally, Panel D shows the case in which allocation concealment is imperfect, so that upcoming allocations cannot be observed, but can be predicted based on the P{E} values. Here the imbalance is severe, 75% females in the active group and 25% females in the control group, but not as severe as it was in the case of no allocation concealment at all. This flow diagram is useful for studying the cause of baseline imbalances (random or selection bias).

At this point the reader may wonder if such a scenario is likely. In fact, the reader would be commended for applying the same scrutiny to this material that this very material asks the reader to apply to the results of randomized trials. But note that at this point, all we set out to do was to establish the theoretical possibility that subversion is possible even in properly randomized trials. There has been no effort, to this point, to comment on the likelihood of this occurring in actual trials. Chapter 3 will present a case that it is, in fact, fairly likely that some randomized trials are subverted through the mechanism presented in this chapter. At that point, the reader may still question if such subversion would have any impact of the conclusions of trials. Chapter 4 will address the impact of selection bias of the results of randomized trials. A critical reader may still ask why we should discuss all of this if nothing can be done about it. The remaining chapters of the book will fill this void, and provide specific measures that can be used to prevent the problem, detect it when it is there, and correct for it when it is found.

3

Evidence of Selection Bias in Randomized Trials

Those looking for a smoking gun may find themselves disappointed with this chapter, and more generally, disappointed with the lack of solid evidence that selection bias exists in randomized trials. Consider, however, where the burden of proof lies. Generally, it is impossible to prove a negative, or the lack of existence of something. It is common, then, to initialize a belief (in the absence of any evidence one way or the other) to the negative, or null hypothesis, and then to be prepared to revise this belief in light of evidence that can disprove it. This is why new treatments need to have their efficacy demonstrated, rather than presumed based on a lack of evidence of ineffectiveness. That the burden of proof be placed on those who would claim the existence of any entity is generally appropriate, and specifically is appropriate when the proof of existence is easier than the proof of absence. But does this argument extend naturally to debate regarding the existence of selection bias in randomized trials? In fact it does not, for reasons to be articulated in this chapter. While the evidence in favor of selection bias in randomized trials may be scant, it will be argued that this evidence is quite strong when compared to either 1) what it *could* be given the forces conspiring against its detection or 2) the evidence of a *lack* of selection bias in randomized trials. It will be argued also that this scant evidence of the existence of selection bias is even stronger when compared to how much evidence would be required to adequately demonstrate that selection bias did not contribute to apparently positive findings in randomized trials, when considering where the burden of proof lies.

Selection Bias and Covariate Imbalances in Randomized Clinical Trials   V. W. Berger
© 2005 John Wiley & Sons, Ltd.

3.1 THE BURDEN OF PROOF REGARDING THE EXISTENCE OF SELECTION BIAS IN RANDOMIZED TRIALS

Even claims by impartial parties with nothing to gain by having others believe this claim can legitimately be challenged. It is even more legitimate, then, to challenge claims of interested parties. As Feigenbaum and Levy (1996) point out,

The pursuit of fortune leads to glaring instances where a researcher may be particularly predisposed to a preferred research outcome. Consider the case of clinical trials of an experimental drug produced by the same pharmaceutical company funding the researcher.

A claim can be further distinguished not only by whether or not the claimant has an interest in having others believe it but also by whether or not the claimant had substantial choice in important aspects of the design, conduct, analysis, and interpretation of the experiment. In a context having nothing to do with selection bias or clinical research, McKay *et al.* (1999) pointed out that when such choice is available,

It is valid to raise the question of whether this lack of tightness in the design of the experiment is at the heart of the result. In precise terms, we ask two questions. Was there enough freedom available in the conduct of the experiment that a small significance level could have been obtained merely by exploiting it? Is there any evidence for that exploitation?

While McKay *et al.* (1999) considered quite different claims, we consider claims of treatment effects in randomized trials, and ask roughly the same questions. That is, we ask if claims of treatment effects in randomized trials can be taken at face value given the current evidence both for and against selection bias. As stated earlier, it is generally impossible to prove a negative, so it is common to initialize a belief to the negative, and hence new treatments need to have their efficacy demonstrated. Likewise, somebody trying to provide a convincing argument regarding the existence of selection bias would have the burden of proof to establish this existence. In a perfect world, the information required to detect selection bias and to quantify its effect in any given trial would be available to all researchers with an interest in testing these hypotheses. However, the reality is that the information required to detect selection bias and to quantify its effect is rarely available from publications (Berger and Weinstein, 2004). Specifically, such key information is generally unavailable to any party other than the sponsor of the trial who 'owns' the raw data. Only this sponsor, then, is in a position to definitively test for the presence or absence of selection bias.

Nance (1998) points out that 'an adverse inference against a party properly arises only if the missing evidence could be presented at reasonable cost by that party'. The Berger–Exner test of selection bias is not costly to conduct, given the availability of the complete data with which to perform it (one caveat being the different choices one could use for $P\{E\}$ when the trial is planned as masked or when block sizes are varied). What, then, is to be made of the fact that the Berger–Exner test of selection bias is not presented in a given trial? In general, a sponsor may be inclined to divulge the results of analyses only if these results support the hypothesis of true superiority. In particular, then, a sponsor may wish to present the results of the Berger–Exner test only if it supports the claim of no selection bias. This last sentence would suggest that the Berger–Exner test is always performed, but presented selectively. In fact, this is probably not the case. In a climate in which the results of analyses are accepted at face value without even having to worry about alternative explanations, including selection bias, it is more likely that failure to present the Berger–Exner test represents failure to conduct it. After all, the perspective of a sponsor might be that no good can come out of looking for selection bias.

If selection bias is found, then this would throw into question any claims of true superiority of the treatment under study. On the other hand, even not finding selection bias after looking for it, and reporting this negative finding, could have an adverse impact on future clinical findings. Such a negative finding would not throw into question claims of true superiority of the treatment presently under study, but it would set a precedent to be followed. That is, if a sponsor were to repeatedly test for selection bias and report that none was found, and then suddenly find selection bias in a given trial and not report anything at all about selection bias for that trial, then this would raise suspicions too. Hence, there is little incentive for a sponsor to agree to subject trial data to rigorous tests of selection bias.

Still, a sponsor who claims a lack of selection bias for a given trial has the data with which to prove the assertion and remove all doubt. Anybody else who expresses uncertainty will necessarily remain in uncertainty until the sponsor decides to reveal the results of these tests of selection bias. As such, we feel that the ability of the sponsor to test for selection bias, coupled with the inability of any other party to do so, shifts the burden of proof towards the sponsor, who must prove that there is no selection bias in a given trial. Yet this is not the only issue to consider in determining where the burden of proof lies. Nance (1998) also states that 'The now conventional understanding of the burden of proof is that the level or weight of the burden of persuasion is determined by the expected utilities associated with correct and incorrect alternative decisions'. It is unclear if a Type I error is always more or always less serious than a Type II error. That is, finding superiority when apparent superiority is an artifact created by selection bias may, in any given circumstance, be deemed more serious or less serious than dismissing true superiority because of a false belief that selection bias was at play.

Because of this uncertainty, it is unclear who would hold the burden of proof in an argument between a sponsor claiming true superiority and a critic claiming nothing but selection bias. Yet these are not the only positions to argue. Apparent superiority of one treatment over another may be attributable to true superiority, selection bias, something else (including other biases), or a combination of the three. It would be appropriate to scrutinize a claim that one, and only one, of these causes is at play, regardless of which cause constitutes the claim. However, uncertainty regarding the cause or causes is the default position, and so this 'lack of a claim' probably constitutes a reasonable default position. The burden of proof falls, then, on whoever would claim that there is certainty, regardless of what that certainty may be.

Such a claim (of certainty) is implicit in a conclusion that one treatment is superior to another, because such a claim implicitly rules out causes other than that the true superiority claimed is the unique cause of the apparent superiority. Again, we acknowledge that a claim of an alternate certainty might require the same burden of proof as the claim of true superiority of a treatment. However, the claim that 'the true superiority claim has not been met' is a weaker claim that can be made with less than a rigid proof of an alternate certainty. It is likely that there would be broad agreement regarding how to interpret clear and convincing evidence of either selection bias or the absence thereof. What remains is the interpretation of the case in which the evidence is not compelling in either direction. The purpose of this chapter is to fulfill the burden of proving not that selection bias is the exclusive cause of apparent treatment effects, but rather that the claims of true superiority being the exclusive cause of apparent treatment effects are premature. As Bailenson and Rips (1996) point out, 'Once one side has fulfilled its burden by presenting convincing evidence, the burden will transfer to the other side of the debate, forcing that side to produce its own evidence or face the prospect of losing the controversy'. So consider, then, an argument between a sponsor claiming true superiority and a critic claiming that true superiority has not been proven, possibly because of the selection bias that the sponsor, in a unique position to do so, did not rule out. In light of the general evidence to be presented in this chapter, it seems rather clear that the burden of proof will transfer to sponsors to offer stronger arguments in support of claims of true superiority.

3.2 INDIRECT POPULATION-LEVEL EVIDENCE THAT SELECTION BIAS EXISTS IN RANDOMIZED TRIALS

Feigenbaum and Levy (1996) noted that, 'Given the asymmetrically high sanctions imposed upon researchers caught in the act of scientific fraud, it is clear that the dominant, low-cost strategy for achieving one's preferred research outcome is to engage in biased methods that fall short of egregious data falsification.' Schulz (1995b) noted that 'practitioners involved in conducting a trial that does not have proper procedures for sequence generation and allocation concealment may find the challenge of deciphering the allocation scheme irresistible'. If both statements are true, then it is not hard to imagine that one who would attempt to predict future allocations would also attempt to exploit this knowledge of future allocations. What patterns would one then expect to see across trials? For one thing, trials without adequate allocation concealment would be expected to produce larger estimates of treatment effects than those with adequate allocation concealment, because those with adequate allocation concealment would not allow for selection bias. Also, there would be more baseline imbalances in unmasked trials (which are susceptible to selection bias) than in masked trials (which are less so). In fact, these patterns have been found repeatedly, by numerous authors.

For example, Chalmers *et al.* (1983) found, among 145 trials classified as masked (57), possibly unmasked (45), and non-random (43), that the rate of significant baseline imbalances (at the 0.05 level) was nearly twice as high for unmasked trials as for masked trials. That is, these baseline imbalances occurred in 14% of the masked trials, 27% of the unmasked trials, and 58% of the non-random trials. Likewise, Schulz *et al.* (1995) reviewed meta-analyses from the Cochrane Collaboration Pregnancy and Childbirth Database, and compared trials with adequate allocation concealment to trials with inadequate or unclear allocation concealment. The results were that the magnitudes of estimated treatment effects were exaggerated by 30% for unclearly concealed trials and by 41% for clearly inadequately concealed trials, relative to those with adequate allocation concealment. This was after adjusting for other measures of quality.

Moher *et al.* (1998) reviewed 127 trials from 11 meta-analyses of interventions for circulatory and digestive disease, mental health, and pregnancy and childbirth. The results were that the magnitudes of estimated treatment effects were exaggerated by 37% for inadequately concealed trials, relative to those with adequate allocation concealment. Kunz and Oxman (1998) found that a lack of randomization and a lack of allocation concealment could increase or decrease the magnitude of the estimated treatment effect, and could even reverse its direction, but on average, the effect was one of increasing the magnitude of the estimated treatment effect.

3.3   DIRECT TRIAL-LEVEL EVIDENCE THAT SELECTION BIAS EXISTS IN RANDOMIZED TRIALS

Specific randomized clinical trials with evidence of selection bias have been described (or at least mentioned) but not identified by Schulz (1995a) and Ivanova *et al.* (2005). Berger and Weinstein (2004) synthesized, but did not attempt to authenticate or refute, published claims that 14 specific randomized clinical trials may have been tainted by selection bias. We now review the key information of these trials in greater detail, and also describe additional examples that have been found since that paper was published. Greenhouse (2003) contends that it 'is obvious that if there are serious imbalances in observable baseline variables, it can only be because clinicians were manipulating patient assignment to a treatment. This by definition should give rise to a selection bias'. We therefore also discuss trials in which severe baseline imbalances were noted, but only if the imbalances were noted to trend in the same direction. We also note that in her Table 3 Reynolds (2004) mentioned a trial in which four pairs of patients had their randomization assignments switched, but insufficient details were provided for this trial to be included in the examples below.

3.3.1 Heparin for myocardial infarction

Carleton *et al.* (1960) discussed a randomized trial in which 125 patients with myocardial infarction were randomized, 60 to heparain and 65 to a control group. The randomization was conducted with sealed envelopes, and Carleton *et al.* (1960) pointed out that the envelopes containing the treatment codes were not numbered consecutively, and there may have been 'prejudice in the selection of therapy by alteration of the sequence of the envelopes. It has been alleged that during the last few weeks or months of this study a few of the envelopes were transilluminated for this purpose'. This is the first publication of which we are aware that mentioned the possibility of the type of selection bias described earlier by Blackwell and Hodges (1957) occurring in an actual randomized trial.

3.3.2 University Group Diabetes Program

The University Group Diabetes Program was designed as a treatment trial, and included tolbutamide and placebo groups. Schor (1971) noted striking baseline imbalances across these treatment groups, and also consistency in the direction of these imbalances. That is, the tolbutamide group had a disproportionate number of patients with cardiovascular risk factors, including elevated cholesterol levels, electrocardiogram abnormalities, obesity, advanced age (over 55 years), a glucose tolerance test value over 750, a history of digitalis, a history of angina, and arterial calcification. Schor (1971) added that

After proper randomization one does not expect to find absolutely similar percentages in both groups [referring to tolbutamide and placebo] for every characteristic. However, one does expect to find certain characteristics which bias the study against tolbutamide to be balanced by other characteristics which bias in favor of tolbutamide. This simply did not happen in this study ... It would appear to any reasonable statistician that for some reason or other the randomization procedure broke down in these three clinics.

Rothman (1977) also noted the need to stratify the analysis of all-cause mortality by age group for this trial. Miettinen and Cook (1981) acknowledged the possibility of selection bias by stating 'by chance *or otherwise*, the tolbutamide series might be of higher risk in terms of its distribution by familiar coronary heart disease risk indicators than the group given placebo' (emphasis added).

That the patients with higher risk tended to be randomized to the active (tolbutamide) group makes this more detailed description consistent with the less detailed description of a trial that was not identified but was mentioned by Ivanova *et al.* (2005). Notice that the suspicion of selection bias was rooted not in a single observed baseline imbalance but rather in the consistency in the directions of the multitude of observed baseline imbalances across the treatment groups. That is, the active group consistently had higher-risk patients.

3.3.3   Talc and mustine for pleural effusions

Numerous baseline imbalances were also found in a randomized comparison of talc to mustine for control of pleural effusions (Fentiman *et al.* 1983). Specifically, Table 3.1 of Altman (1985) demonstrates severe baseline imbalances in age (means of 50.3 years for mustine vs. 55.3 years for talc), stage 1 or 2 (52% for mustine vs. 74% for talc), mean interval between breast cancer diagnosis and effusion diagnosis (33.1 months for mustine vs. 60.4 months for talc), and proportion post-menopausal (43% for mustine vs. 74% for talc). These imbalances were quite large in magnitude, yet did not reach statistical significance at the customary level (0.05) because

Table 3.1 Some baseline characteristics of patients in controlled trial of Fentiman *et al.* (1983) reproduced from The Statistician, Vol. 34, No. 1, Statistics in Health (1985), 125–136

	Treatment	
	Mustine ($n = 23$)	Talc ($n = 23$)
Mean age (SE)	50.3 (1.5)	55.3 (2.2)
stage 1 or 2	52%	74%
stage 3 or 4	48%	26%
Mean interval in months (SE) between breast cancer diagnosis and effusion diagnosis	33.1 (6.2)	60.4 (13.1)
Postmenopausal	43%	74%

of the small sample size (23 patients randomized to each treatment group).

3.3.4 Tonsillectomy for recurrent throat infection in children

The imbalances did reach significance in a randomized trial of tonsillectomy for recurrent throat infection in children (Paradise *et al.*, 1984), with $p = 0.0309$ and $p = 0.0076$, respectively, by the two-sided Smirnov test (Berger *et al.* 1998), for history of episodes of throat infection and parents' socioeconomic status. This was an unmasked trial, and blocks of fixed size 4 were used, so the trial was certainly susceptible to selection bias.

3.3.5 Oxytocin and amniotomy for induction of labor

Oxytocin was compared to amniotomy for induction of labor in 223 women who were allocated based on even or uneven dates of birth (Bakos and Backstrom, 1987). Clearly this trial does not qualify as truly randomized (Berger and Bears, 2003). Nevertheless, it is labeled as a randomized trial, and a casual reading would not reveal that in fact it was not a randomized trial, so we include it in this chapter. A reanalysis

> hypothesized that clinicians knowing in advance the method of induction of labor to be used for each woman would be influenced in their decision to use induction at all (enroll the woman in the trial). It was demonstrated that obstetricians were very reluctant to induce labor with amniotomy in a woman born on an uneven date when she had an unfavorable cervix (low Bishop score). Thus, randomization by date of birth was an unsatisfactory method in this case, because it produced selection bias at trial entry. (Villar and Carroli, 1996)

More specifically,

> women with a Bishop score of three or less were four times more likely to be entered into the trial if they were born on an even date (allocated to the oxytocin group) than if they were born on an uneven date (allocated to the amniotomy group). Of 35 such women, only seven (20%) were allocated to amniotomy and 28 to oxytocin infusion, a difference that is statistically significant at $p < 0.0005(\chi^2 = 11.43)$. If women with an unfavorable cervix had been scheduled for induction with the same frequency, irrespective of whether they were born on even or on uneven dates, the likelihood that they would be divided among the amniotomy and oxytocin groups as found . . . is less than one in 4,000 . . . The null hypothesis that the Bishop score before induction did not influence the decision to assign women to either the amniotomy or the oxytocin group was thus rejected at $p < 0.00025$. (Keirse, 1988)

3.3.6   Western Washington Intracoronary Streptokinase Trial

In the Western Washington Intracoronary Streptokinase Trial, comparing intracoronary streptokinase to standard medical attention for the treatment of acute myocardial infarction, 250 patients were randomized, 134 to streptokinase and 116 to control, but one patient was mistakenly assigned to streptokinase (Hallstrom and Davis, 1988). Given the planned 1:1 allocation ratio, the probability of observing a difference of at least 18 in the group sizes is 0.004 (Hallstrom and Davis, 1988). The probability of observing even the corrected difference of 16 or larger is still very low (0.009). The statisticians involved in the study were 'particularly concerned in verifying that the randomization process had been carried out as planned', as these values suggest the possibility of basing treatment assignments on patient characteristics, a form of selection bias (Hallstrom and Davis, 1988).

3.3.7 RSV immune globulin in infants and young children with respiratory syncytial virus

Groothuis *et al.* (1993) did not fully describe the method by which treatments were assigned in the study of RSV immune globulin in infants and young children with respiratory syncytial virus. However, it is clear that randomization was unmasked, and was developed and performed separately by each center (Ellenberg *et al.*, 1994). Any restrictions on the randomization would then allow for prediction of future assignments. The possibility that this information was used to selectively enroll healthier or sicker patients across groups was raised (Ellenberg *et al.*, 1994), as 'the person maintaining the list [at each center] would have been aware of the upcoming treatment assignments'. The methods they used for randomization 'did not protect against such influences, conscious or unconscious' (Ellenberg *et al.*, 1994).

3.3.8 A trial to assess episiotomy

A trial to assess episiotomy was to have randomized patients to either liberal or restrictive use of episiotomy, but this trial was 'affected by physician noncompliance with the randomly assigned therapy' (Schulz, 1995b). That is, it appeared that some physicians assigned episiotomy to certain patients even when it was not the treatment indicated by the randomization. While assigning a treatment deterministically does not require advance knowledge of upcoming allocations, and hence is not exactly the type of selection bias we discuss, it is still a form of treatment allocation based on patient characteristics, and introduces selection bias. Moreover, Schulz (1995b) noted that 'The physicians who viewed episiotomy more favorably decided more frequently not to randomly assign certain participants, who had been enrolled, to a study arm', and asked if they had knowledge of the next treatment to be allocated.

3.3.9   Canadian National Breast Cancer Screening Study

Cohen *et al.* (1996) described how randomization was conducted in the Canadian National Breast Screening Study:

Women were interviewed, filled out an entry questionnaire and consent form, and had a physical examination of the breasts. Only after this initial review was a woman randomized to the mammography or alternative arms of the trial. Randomization was carried out as follows: Women meeting eligibility criteria were entered onto a list preprinted with identification number, trial arm designation, and space for the participant's name . . . Because of the nonblinded nature of this process, it was possible for NBSS staff to preferentially assign some women to the mammography arm by putting their names opposite a mammography designation rather than in the next blank space.

Bailar and MacMahon (1997) noted that, as stated by Cohen *et al.* (1996), women were randomized only *after* the clinical examination, but that this was not the case in one center (center 03). The significance of this is apparent only when one considers the lack of allocation concealment.

That is, 'after each subject had been examined by a nurse (or a physician, in Quebec), but before randomization took place, the coordinator knew the group to which the next subject would be allocated. In principle, subversion of the sequence of events prescribed by the investigators could thus have taken place' (Boyd, 1997). Bailar and MacMahon (1997) also noted that with the exception of center 03,

the nurses (and probably also the coordinators) were aware of the findings of the clinical examination when the allocation was made. Herein lies the basis of the charge that examiners who thought that a woman should or should not have a mammogram, because of findings at clinical examination or personal information obtained during the examination (e.g., risk factors for breast cancer), may have compromised the randomization.

Boyd (1997) tabulated (in his Table 3.2) some baseline imbalances across the groups, including prior health claims for breast cancer in women aged 40–59 (8 vs. 1, $p = 0.05$), alterations of names in allocation books in women aged 40–49 (95 vs. 65, $p = 0.01$), and advanced breast cancer detected at baseline by physical examination

Table 3.2   Frequency of selected events in the Canadian National Breast Screening Study, by study arm reproduced by permission of Canadian Medical Association

| Event | Age | Study arm; no. of subjects | | p value |
		Mammography	Usual care	
Prior health claim for breast cancer[10]	40–59	8	1	0.05†
Alteration of name in allocation book*	40–49	95	65	0.01
Advanced breast cancer detected at baseline by physical examination[7–9]	40–49	17	5	0.003
Death from breast cancer in 7-year follow-up period[8]	40–49	38	28	0.27†

* Data from the review by Bailar and MacMahon (see pages 193 to 199).
† Calculated by the χ^2 test.

in women aged 40–49 (17 vs. 5, $p = 0.003$). Boyd (1997) also pointed out that

great deviousness would not have been required to achieve a particular allocation . . . Suppose a subject wished to be allocated to the mammography arm? . . . If the next allocation was to the control arm instead, the subject's name could have been entered onto the line with the next mammography allocation, leaving a gap in the allocation book, or she could have been advised to wait until the line for the desired arm was the next to be filled. In either case, it is unlikely that much time would have elapsed before a mammography allocation came up or a gap on the list was filled. Fifteen NBSS centers randomly assigned 90 000 women over five years or less: They must have been busy places.

3.3.10   Surgical trial

In a surgical trial conducted at 23 centers, some centers used the sealed envelope system and others used the centralized telephone system (Kennedy and Grant, 1997). The median age of patients randomized to the experimental treatment was considerably lower than those in the conventional treatment group (59 vs. 63 years, $p < 0.01$) when envelopes were used. For three clinicians there were even larger age imbalances across treatment groups (57 vs. 72 years, $p < 0.01$; 33 vs. 69 years, $p < 0.001$; 47 vs. 72 years, $p = 0.03$). These imbalances were not observed when using the telephone system (Kennedy and Grant, 1997), so the implication is that these imbalances resulted from the use of the sealed envelope system itself. The mechanism would then be through the ability to observe upcoming allocations to be made, or selection bias.

3.3.11   Lifestyle Heart Trial

The Lifestyle Heart Trial was conducted to assess the progression of coronary atherosclerosis achieved through drastic lifestyle changes without lipid-lowering drugs (Ornish *et al.*, 1998). Of 193 patients, 93 remained eligible after a quantitative coronary angiography. Of these, 53 were randomized (randomization details are unclear, but the study appears to have been unmasked) into the experimental group, and 40 to the control group. However, only 28 patients randomized to the experimental group and 20 patients randomized to the control group agreed to participate in the study. Not all patients who agreed to participate reported their data, and therefore the analyses presented were based on only 20 patients in the experimental group and on only 15 patients in the control group. When comparing these two groups, there was a baseline imbalance in gender – all three women were in the control group, $p = 0.07$, which is a fairly low p-value considering the small sample sizes of the groups being compared (see Section 3.3.3).

3.3.12   Coronary Artery Surgery Study

The Coronary Artery Surgery Study (CASS) compared coronary by-pass surgery to medical therapy. Assignments were made via tele-phone communication from the coordinating center. Of the 2099 randomizable patients, 780 agreed to participate. The randomized and randomizable groups (that is, those patients who could have been randomized but were not) differed in the degree of baseline coronary artery disease (CASS Investigators, 1984). Berger and Exner (1999) pointed out that within the medical therapy group there was *more* extensive baseline coronary artery disease among randomized than randomizable patients, whereas within the surgery group there was *less* extensive coronary artery disease among randomized than randomizable patients. Furthermore, Berger and Exner (1999) commented on the implications this pattern has for selection bias. One could describe this pattern as healthier patients apparently being selected for the surgery group and turned away for the medical group, whereas sicker patients seemed to be selected for the medical group and turned away for the surgery group.

3.3.13   Etanercept for children with juvenile rheumatoid arthritis

In a study of etanercept for children with juvenile rheumatoid arthritis (Lovell *et al.*, 2000), patients in the etanercept group were younger (mean 8.9 years vs. 12.2 years, $p = 0.0026$), less likely to be Caucasian (56% vs. 88%, $p = 0.022$), and of lower weight (mean 34 kg vs. 43 kg, $p = 0.027$) than patients in the placebo group. Lovell *et al.* (2000) make no attempt to explain these baseline differences beyond stating that 'The unequal randomization did not affect the study results', but the Food and Drug Administration statistical review (Berger, 1999), which was the source for these baseline p-values, reveals several other issues as well.

For one thing, there was a three-month open-label run-in on etanercept. Because it is easier to discern similarity to, or difference from, that which has already been experienced than it is to unmask an assignment when neither treatment has been previously experienced (Leber and Davis, 1998), this run-in increases the likelihood of unmasking treatment allocations. Also, the randomization used blocks of size 2, the worst situation for selection bias (Proschan, 1994). Worse still, corresponding blocks in the two strata within a center (stratification was based on study center and the number of active joints being no greater than 2 or greater than 2) were mirror images of each other. For example, in site 514, the first patient in the 'few active joints' stratum was to receive placebo. This implies that the first patient in the 'more than two active joints' stratum within site 514 necessarily was to receive etanercept. In essence, unmasking one allocation leads to perfect advance knowledge of three other allocations.

Beyond the baseline imbalances, four patients were also randomized from the wrong stratum. In three of these cases, it was foreseeable that the treatment received would likely be affected. That is, in one case, the patient had more than two active joints, and the next allocation in this stratum was to etanercept with certainty. Yet the patient was allocated from the 'few active joints' stratum, with certainty of receiving placebo. The next mistaken allocation was of a patient with few active joints, and etanercept would have been received with certainty. Instead, this patient was randomized from the 'many active joints' stratum, which had a fifty–fifty chance for either treatment (because this was the first allocation in the block, and the mirror image had not yet been revealed). As it turned out, etanercept was allocated anyway. The final mistake (there were actually four, but we detail only the three for which it was foreseeable that the treatment received would likely be affected) affected a patient with few active joints, and placebo would have been received with certainty. Instead, this patient was randomized from the 'many active joints' stratum, which had a fifty–fifty chance for either treatment (because this was the first allocation in the block, and the mirror image had not yet been revealed). As it turned out, etanercept was allocated. In two cases, then, the treatment received was actually reversed by these mistakes. Neither this, nor the fact that some patients were randomized out of order, was mentioned in the publication (Lovell *et al.*, 2000).

3.3.14   Edinburgh Randomized Trial of Breast-Cancer Screening

The Edinburgh Randomized Trial of Breast-Cancer Screening, begun in 1978, used cluster randomization, and had a baseline imbalance in socioeconomic status. Alexander *et al.* (1999) connected the imbalance to the use of cluster randomization, stating that, 'As a result of the cluster randomisation, there was bias between the two groups, women in the control group having higher all-cause mortality rates and lower socioeconomic status (SES) than those randomly assigned to intervention'. That cluster randomization can, by its very nature, interfere with allocation concealment was pointed out by Jordhoy *et al.* (2002); see Section 3.3.24. In the bottom row of their Table 3.3, under the column for the Edinburgh Trial, Humphrey *et al.* (2002) wrote 'Allocation concealment not described; significantly lower SES and higher all-cause mortality in control group suggest inadequate randomization'.

3.3.15   Captopril Prevention Project

In the Captopril Prevention Project, 10 985 patients were followed for a mean of 6.1 years after having been randomized to captopril or conventional antihypertensive therapy, and sealed envelopes were to be used (Psaty *et al.*, 2000). Peto (1999) stated that the baseline differences between the two treatment groups in

height, weight, systolic, and diastolic blood pressure (with respective p-values of 10^{-4}, 10^{-3}, 10^{-8}, and 10^{-18}) show that the process of randomisation by sealed numbered envelopes was frequently violated. Presumably, at some centres those responsible for entering patients sometimes unsealed the envelopes before the next patient was formally entered, and then let knowledge of what the next treatment would be influence their decision as to whether that patient should be entered and assigned that foreknown treatment.

3.3.16   Göteborg (Swedish) Mammography Trial

Gotzsche and Olsen (2000b) pointed out that 'In the trial from Göteborg, the imbalance in age was small but the authors note that 28% of 1655 women surveyed in the study group reported having undergone a mammogram before the trial, compared with 51% of 1641 controls. This suggests that the groups were not comparable at baseline ($p = 8 \times 10^{-42}$)'.

3.3.17   HIP Mammography Trial

Regarding the HIP (Health Insurance Plan) Mammography Trial, begun in New York in the 1960s, Kolata (2002) stated that

researchers began by randomly assigning women to have mammograms or not. But they also decided that they did not want to include women who already had breast cancer. So after the women were assigned, they dropped

Table 3.3 Controlled Trials of Mammography and Clinical Breast Examination* reproduced from the Annals of Internal Medicine, www.annals.org, Vol. 137, No. 5 (Part 1), E-349

Variable	Trial (Reference)							
	HIP (19)	CNBSS-1 (13)	CNBSS-2 (13, 20)	Edinburgh (18)	Gothenburg (14, 23)	Stockholm (17)	Malmö (15)	Swedish Two-County Trial (16)
Description								
Year study began	1963	1980	1980	1978	1982	1981	1976–1978	1977
Setting or population	New York health plan members	15 centers in Canada, self-selected participants	15 centers in Canada, self-selected participants	All women aged 45–64 y from 87 general practices in Edinburgh	Entire female population, born between 1923–1944, of one Swedish town	Residents of south-east greater Stockholm, Sweden	All women born between 1927–1945 living in Malmö, Sweden	From Ostergotland (E-County) and Kopparberg (W-County)
Age at enrollment, y	40–64	40–49	50–59	45–64	39–59	40–64	45–70	40–74

interventions

Method of randomization	Age- and family size–stratified pairs of women randomly assigned individually by drawing from a list	Blocks (stratified by center and 5-year age group) after CBE	Cluster, based on general practitioner practices	Cluster, based on day of birth for 1923–1935 cohort (18%), by individual for 1936–1944 cohort (82%)	Individual, by day of month; ratio of screening to control group, 2:1	Individual, within birth year	Cluster, based on geographic units; blocks designed to be demographically homogeneous
Study groups	Mammography + CBE vs. usual care	Mammography + CBE vs. CBE (all women pre-screened and instructed in BSE)	Mammography + CBE vs. usual care	Mammography vs. usual care; controls offered screening after year 5, completed screening at approximately year 7	Mammography vs. usual care; controls offered screening after year 5	Mammography vs. usual care; controls offered screening after year 14	Mammography vs. usual care; controls offered screening after year 7

Continued

Table 3.3 (*Continued*)

	Trial (Reference)							
Variable	HIP (19)	CNBSS-1 (13)	CNBSS-2 (13, 20)	Edinburgh (18)	Gothenburg (14, 23)	Stockholm (17)	Malmö (15)	Swedish Two-County Trial (16)
Screening protocol								
Interval, mo	12	12	12	24	18	24–28	18–24	24–33
Rounds, n	4	4–5	4–5	4	5	2	9	3
Views, n	2	2	2	2 (1)	2 (1)	1	2 (1)	1
Participants, n								
Study group	30 239	25 214	19 711	28 628	20 724	40 318	21 088	77 080
Control group	30 256	25 216	19 694	26 015	28 809	19 943	21 195	55 985
Longest follow-up by 2002, y	18	13	13	14	12†	11.4†	11–13	20
							15.5†	15.5†

Trial quality
Assembly of comparable groups
Allocation concealment and baseline groups

Use of lists and pairs made subversion possible. More menopausal women and women with previous breast lumps in a sample of controls; more education in the screened group	Use of lists and blocks made subversion possible 17 in women in mammography group vs. 5 in control group had tumors with 4 nodes on initial screening	Allocation concealment not described; significantly lower SES and higher all-cause mortality in control group suggest inadequate randomization	Use of lists and blocks made subversion possible	Allocation concealment not described	Allocation concealment not described	Allocation concealment not described	Allocation concealment not described; intervention women slightly older than controls

Continued

Table 3.3 (*Continued*)

Variable	HIP (19)	CNBSS-1 (13)	CNBSS-2 (13, 20)	Edinburgh (18)	Gothenburg (14, 23)	Stockholm (17)	Malmö (15)	Swedish Two-County Trial (16)
					Trial (Reference)			
Relative risk for all-cause mortality (screened vs. control group)	0.98	1.02	1.06	0.8 (*statistically significant*)	0.98	NR	0.99	1
Maintenance of comparable groups								
Screening attendance	Round 1, 67%; round 2, 54%; round 3, 50%; round 4, 46%	Round 1 100%; rounds 2 and 4.85%–89%	Round 1, 100%; round 2, 90.4%, round 5, 86.5%	Round 1, 61%; round 7, 44%	Round 1, 85%; rounds 2–5, 75%–78%; control group, 66%	Round 1, 81%; round 2, 81%; control group, 77%	Round 1, 74%; rounds 2–5, 70%; control group, ???	Round 1, 89%; round 2, 83%; round 3, 84%; control group, ???

Contamination, %	Unknown, probably small	16	Not reported	20	Not reported	25	13
Post randomization exclusions	Yes	No	Yes	Yes	Yes	Yes	Yes
Validity of outcome assessment							
Deaths included in analysis (follow-up vs. evaluation method)	*Breast cancer deaths diagnosed within 7 years of follow-up* Follow-up method	Follow-up method	Follow-up method	Follow-up method and evaluation method	One fewer death in screening group included in 1997 results	*Initially, all four trials used the evaluation method of analysis (breast cancer cases diagnosed after screening period were excluded from count of breast cancer deaths), but this was corrected in reanalyses of the data in 1993 and in 2002. Control screening was delayed relative to the last screen in the mammography groups, resulting in bias because more cases of cancer were included in the control groups than in the intervention groups.*	

Continued

Table 3.3 (Continued)

			Trial (Reference)				
Variable	HIP (19)	CNBSS-1 (13)	CNBSS-2 (13, 20)	Edinburgh (18)	Gothenburg (14, 23)	Stockholm (17)	Malmö (15)	Swedish Two-County Trial (16)
Method for verifying breast cancer deaths	Blinded review of the death certificate and medical records; *unclear how deaths were selected for review*	Blinded review of all deaths of women known to have breast cancer whose death certificates mentioned liver, lung, or colon cancer or unknown primary, or whose medical records raised a question of breast cancer		All deaths, with breast cancer deaths diagnosed within 14 years of follow-up; *not masked*	In the 1993 analysis, an independent panel used an explicit protocol to perform blinded assessment of cause of death			

Analysis method					
Intention-to-treat analysis; completeness of reporting‡	*Did not provide relative risk, confidence intervals, or P values in recent report; estimated the number of participants*	Appropriate	Appropriate	—	Sample sizes differed for different publications because different methods were used to estimate the size of the underlying population.

(Continued)

Table 3.3 (*Continued*)

				Trial (Reference)				
Variable	HIP (19)	CNBSS-1 (13)	CNBSS-2 (13, 20)	Edinburgh (18)	Gothenburg (14, 23)	Stockholm (17)	Malmö (15)	Swedish Two-County Trial (16)
External validity	Poor mammography technique; only a third of cancer cases found by mammography alone	Many women with screening abnormalities (especially on CBE) were "deemed not to require a diagnostic procedure," potentially reducing the sensitivity of screening		–	19% of controls and 13% of study women had had mammography in the 2 years before the study	25% of all women entering the study had had mammography	–	In the age group of 40–49 y, 3 women died after being invited to screening and 1 died before invitation but after randomization
Grade								
USPSTF internal validity	Fair	Fair or better	Fair or better	Poor	Fair	Fair	Fair	Fair

* Italic type indicates aspects of the design or conduct of the trials that influenced the quality rating. BSE = breast self-examination; CBE = clinical breast examination; CNBSS = Canadian National Breast Screening Study; HIP = Health Insurance Plan of Greater New York; NR = not reported; USPSTF = U.S. Preventive Services Task Force.

† Most recent results for age 40 to 49 years, if different.

‡ All studies were analyzed by using intention-to-treat methods.

women who, they later realized from looking at medical records, had had cancer. About 1100 ended up being dropped – some 800 from the mammography group and about 300 from the control group. Critics of the study wonder why so many more women in the screening group turned out to have had a diagnosis of breast cancer before the study began. In theory, they say, the numbers should have been roughly equal.

Humphrey *et al.* (2002) added that the 'Use of lists and pairs made subversion possible. More menopausal women and women with previous breast lumps in a sample of controls; more education in the screened group.'

3.3.18 Hypertension Detection and Follow-Up Program

The Hypertension Detection and Follow-Up Program (HDFP) used sealed envelopes for the randomization which, as noted by Psaty *et al.* (2000), is subject to manipulation. In fact, the randomization 'was tampered with at one clinic and as a result, 446 participants from that clinic were excluded from HDFP analyses'.

3.3.19 Randomized trial to prevent vertical transmission of HIV-1

Another trial with block randomization, sealed envelopes (which, as we have seen in Section 3.3.18, is subject to manipulation), and a (nearly significant) baseline imbalance is the trial described by Hughes and Richardson (2000). Specifically, women were randomized to either formula-feed (212) or breast-feed (213) their infants. After 24 exclusions (16 in the breast-feeding arm and only eight in the formula arm), there were 401 infants available for analysis. The exclusions were for stillbirths, second-born twins, and mothers lost before delivery, and the one-sided p-value comparing the rates of these exclusions across arms is $p = 0.0716$.

3.3.20 Effectiveness trial of a diagnostic test

Swingler and Zwarenstein (2000) described problems with the randomization and allocation concealment of an effectiveness trial of a

Table 3.4   Problems with the sequencing and allocation of potentially eligible patients, and the fate of envelopes of excluded patients reprinted from Journal of Clinical Epidemiology, 53 (2000) 704, with permission from Elsevier

Registration of potentially eligible patients before randomization	
Potentially eligible patients	581
Deletions or alterations in register	0
Enrolled out of chronological sequence	9 (1.5% of 581)
Change to radiograph allocation	2
Change to non-radiograph allocation	4
No change in allocation	3
Exclusions before randomization	59 (10.2% of 581)
Allocation envelopes improperly handled (excluded patients)	16 (27.1% of 59)
Opened	7
Lost	9

diagnostic test. Their Table 3.4 shows that nine enrollments were out of chronological sequence, and that six of these nine resulted in a change in the allocation. Nine allocation envelopes were also lost, and seven were opened, for a total of 16 that were mishandled. Of these 16, four had radiograph allocations, and 12 had non-radiograph allocations ($p = 0.05$). The authors noted that the

preponderance of non-radiograph allocations among the 16 excluded cases with open or lost envelopes is unlikely to have happened by chance ($p = 0.05$). This imbalance in allocation suggests that some subjects may have been excluded from the study because of non-radiograph allocations. The high proportion of such patients with opened envelopes and non-radiograph allocations who received a radiograph, compared with patients with lost envelopes, supports this hypothesis.

3.3.21   South African trial of high-dose chemotherapy for metastatic breast cancer

Weiss *et al.* (2001) audited a randomized South African trial of high-dose chemotherapy (HDC) for metastatic breast cancer (Bezwoda *et al.*, 1995), and described a lack of allocation concealment, and linked this to questioning the integrity of the randomization process. Specifically, Weiss *et al.* (2001) stated that the lead investigator had a logbook listing the allocation sequence, but replacing the identity of the treatments with the generic 'A' and 'B', and went on to point out that, 'Of the first 24 patients entered, only six were designated to have received HDC. Then, of the last 27 patients, only six were designated to have received the CNV (cyclophosphamide, mitoxantrone, and vincristine) regimen. It is unlikely that this sequence of treatment assignments could have occurred if the study were truly randomized.'

3.3.22 Randomized study of a culturally sensitive AIDS education program

In a randomized study of a culturally sensitive AIDS education program (Stevenson and Davis, 1994), Marcus (2001) hypothesized that 'subjects with lower baseline knowledge scores . . . may have been channelled into the treatment group', because the treatment group had significantly lower baseline AIDS knowledge scores (39.89 vs. 36.72 on a 52 question test, $p = 0.005$).

3.3.23 Runaway Youth Study

Song *et al.* (2001) described the Runaway Youth Study (aimed at preventing HIV transmission among runaway youths in New York City), and its baseline imbalances. The randomization was performed in clusters, with two shelters (with 167 youths) allocated to the intervention group (staff training and a series of interactive group sessions), and two shelters (with 144 youths) allocated to the control treatment (no specialized intervention). Song *et al.* (2001) stated:

Not surprisingly given that randomization occurred at the level of the shelter rather than by participant, the intervention and control groups were not comparable at recruitment. As shown in Table 1, several differences emerged on sociodemographic and substance-use characteristics; in all, significant differences were seen on nine of 45 baseline characteristics. For example, youths in the control shelter were on average roughly one year older, had a higher school dropout rate, and had more extensive alcohol and drug use.

As Table 3.5 of Song *et al.* (2001) shows, some of these imbalances had strikingly low p-values. This includes age (16.2 years in the control group vs. 15.1 years in the intervention group, $p < 0.001$), reason for leaving home ($p = 0.004$), school dropout (48.3% in the control group vs. 32.3% in the intervention group, $p = 0.004$), lifetime

Table 3.5   Selected baseline characteristics between the intervention and the control groups reprinted from Handling Baseline Differences and Missing Items in a longitudinal Study of HIV Risk Among Runaway Youths, Song *et al.*, Health Services & Outcomes Research Methodology 2:317–329, 2001, with kind permission of Springer Science and Business Media

	Sample size	Control ($n = 144$)	Intervention ($n = 167$)	p-value
Male	311	47.9%	53.3%	0.34
Age (SD)	311	16.2(1.51)	15.1(1.74)	<0.001
Ethnicity	306			
African American		54.6%	62.4%	0.28
Hispanic		27.7%	25.5%	
White/Others		17.7%	12.1%	
Reason left home	309			
Other people had problems		19.4%	12.7%	0.004
Someone forced out		21.5%	13.9%	
Neglect/abused		13.9%	16.4%	
Getting into trouble		16.7%	8.5%	
Trouble with parents		13.2%	24.9%	
Other reasons		15.3%	23.6%	
Number of times runaway from home (SD)	275	2.56(1.85)	2.54(1.95)	0.93
Number of times forced out (SD)	207	1.22(1.78)	0.89(1.47)	0.14
School dropout	307	48.3%	32.3%	0.004
Suicide				
Suicide attempt	296	26.1%	29.6%	0.50
Family suicide	300	34.8%	45.7%	0.06
Conduct disorder (SD)	310	1.10(0.73)	1.02(0.79)	0.37
Lifetime sexual behavior				
Abstinent	308	16.7%	18.3%	0.71
Lifetime drug and alcohol				
Alcohol use	307	78.5%	63.8%	0.005
Drug use	305	52.1%	36.7%	0.007
Sexual Behavior for the past 3 months				
# of partners (SD)	308	1.96(3.34)	1.95(2.96)	0.97
# of unprotected sexual acts (SD)	308	10.34(20.61)	7.77(16.55)	0.23
Drug and alcohol for the past 3 months				
Alcohol	301	62.4%	39.4%	<0.001
Marijuana	303	31.5%	17.5%	0.005
Hard drug	307	11.1%	6.1%	0.12
Number of drugs (SD)	307	0.43(0.68)	0.26(0.57)	0.02

alcohol use (78.5% in the control group vs. 63.8% in the intervention group, $p = 0.005$), lifetime drug use (52.1% in the control group vs. 36.7% in the intervention group, $p = 0.007$), recent alcohol use (62.4% in the control group vs. 39.4% in the intervention group, $p < 0.001$), and recent marijuana use (31.5% in the control group vs. 17.5% in the intervention group, $p = 0.005$).

3.3.24 Cluster randomized trial of palliative care

Jordhoy *et al.* (2002) discussed a cluster randomized trial of palliative care conducted at the Palliative Medicine Unit of Trondheim University Hospital in Norway, and noted that allocation concealment may be impossible when cluster randomization is used, a point that was hinted at by Song *et al.* (2001); see Section 3.3.23. The trial presently considered used three pairs of clusters, and randomly assigned one cluster per pair to the intervention, with the other cluster receiving the control; see Table 3.6 of Jordhoy *et al.* (2002).

In their Table 3.7, Jordhoy *et al.* (2002) display baseline imbalances, including 45% of the intervention group (106/235) vs. 28% of the control group (56/199) living in a villa ($p = 0.001$), 23% of the intervention group (54/235) vs. 40% of the control group (80/199) living in an apartment ($p < 0.001$), and 11% of the intervention group (26/235) vs. 23% of the control group (45/199) receiving home care nursing ($p = 0.002$). Additional baseline imbalances (medical, as opposed to sociodemographic) are displayed in Table 3.8 of Jordhoy *et al.* (2002), including 73/235 intervention patients (31%) vs. 89/199 control patients (45%) with specific types of cancer ($p = 0.005$). Jordhoy *et al.* (2002) noted that:

The individual patient results suggested that diagnosis was not randomly distributed across the two groups, and the cluster-adjusted analysis confirmed that there was an imbalance even when allowing for the clustering due to healthcare districts and trial design. Thus, the statistical results supported our suspicion of biased selection . . . we believe it unlikely that the imbalance on diagnoses could be related to a real difference in cancer incidence among the clusters . . . Since patients' allocation was predictable by their address, it seemed reasonable that such factors affected the two arms differently. Hence, based on the statistics, the magnitude and direction of the imbalance, and our knowledge of the local health care system, selection bias was the most obvious explanation.

Table 3.6 Participating communities and community health care districts by randomized clusters[*] reprinted by permission of Edward Arnold (Publishers) Ltd

District	Urban (U), rural (R)	Inhabitants, n	Inhabitants >60 years		Cluster	Pair, n	randomization allocation
			$n^{\dagger}$	%			
Byåsen	U	28 945	6052	21	Byåsen/Heimdal	I	Intervention
Heimdal	U	25 655	2824	11			
Sentrum	U	29 190	7755	27	Sentrum/Saupstad	I	Control
Saupstad	U	13 267	1185	9			
Nardo	U	17 930	3686	21	Nardo	II	Intervention
Strinda	U	24 851	4928	20	Strinda	II	Control
Malvik	R	10 329	1523	15	Malvik	III	Intervention
Melhus	R	12 706	2159	17	Melhus	III	Control

[*] Randomization was based on independent health care districts. Clusters were defined and stratified into three pairs according to number of inhabitants above 60 years of age and whether they represented rural or urban area.
[†] Official numbers provided for Trondheim (1/1–93) by Statistisk sentralbyrå (Statistics Norway) and for Malvik and Melhus (1/1–94) by the Community Registration Office.

Table 3.7 Patients' sociodemographic characteristics at baseline according to treatment group and cluster reprinted by permission of Edward Arnold (Publishers) Ltd

Patients' characteristics*	I (intervention) n	I (intervention) %	C (control) n	C (control) %	Pair I I %	Pair I C %	Pair II I %	Pair II C %	Pair III I %	Pair III C %	Significance of difference between treatment groups (P value)†
Female	103	44	101	51	45	52	46	52	33	39	0.15
Male	132	56	98	49	55	48	54	48	67	61	
Age, median years (min–max)	70	38–90	69	37–93	71	72	69	68	72	67	0.93
Married/cohab	158	67	117	59	71	53	62	66	63	72	0.07
Widowed/divorced/unmarried	77	33	82	41	29	47	38	34	37	28	
Living alone	70	30	71	36	25	42	35	28	38	22	0.21
Living with spouseChildren	157	67	116	58	70	52	62	66	62	72	0.07
Living with others	8	3	12	6	5	6	3	6	0	6	0.22
Living in a villa‡	106	45	56	28	51	16	27	32	71	89	0.001
Living in an apartment	54	23	80	40	19	54	36	26	0	0	<0.001
Living in a semi-attached house	75	32	61	31	30	28	36	42	29	6	0.830
Having less than 7 years of education	93	40	67	34	31	32	49	32	54	50	0.19

(Continued)

Table 3.7 (Continued)

Patients' characteristics*	I (intervention)		C (control)		Pair I		Pair II		Pair III		Significance of difference between treatment groups (P value)[†]
	n	%	n	%	I %	C %	I %	C %	I %	C %	
Having 8–10 years of education	72	31	77	39	32	41	34	37	13	33	0.10
Having 11–12 years of education	40	17	22	11	21	9	10	15	17	11	0.09
Having more than 13 years of education	30	13	33	17	16	19	7	15	17	6	0.27
Having relatives in same/neighbour community	214	91	179	90	90	91	94	86	88	100	0.72
Having access to informal help	187	80	140	70	81	72	73	65	96	78	0.04
Receiving informal help	139	59	110	55	60	59	61	49	46	56	0.42
Receiving home care nursing	26	11	45	23	8	22	10	26	29	17	0.002
Number of patients	235		199		134	116	77	65	24	18	

* All characteristics represent individual variables for which the difference in distribution between treatment groups was tested; for these analyses, categorical variables were coded as having the characteristic = 1, not having the characteristic = 0.
[†] P values for the significance of difference were obtained by bootstrap estimation to fit regression models using the variable to the left as the dependent variable.
[‡] Information on two control patients was missing.

Table 3.8 Patients' baseline medical characteristics according to treatment group reprinted by permission of Edward Arnold (Publishers) Ltd

	Intervention		Control		Significance of difference between treatment groups (P value)†
	n	%	n	%	
Having chronic disorders other than cancer	110	47	87	44	0.53
Having disabilities (not related to cancer)	116	49	98	49	0.98
Having cancer origin group I*	137	58	96	48	0.04
Gastrointestinal	107		74		
Lung	30		22		
Having cancer origin group II*	73	31	89	45	0.005
Breast and female genitals	26		41		
Prostate and male genitals	21		20		
Kidney/vesica	16		13		
Lymphomas	4		9		
Skin	6		6		
Having cancer origin group III*	25	11	14	7	0.42
Unknown origin	13		10		
Others	12		4		
Median weeks from diagnosis to inclusion (min–max)‡	29	0–1469	57	–1–1087	0.25
Having Karnofsky status below or similar to 70	88	37	82	41	0.43
Having Karnofsky status above 70	147	63	117	59	
Having distant metastasis	193	82	150	76	0.10

†P values for the significance of difference between treatment groups as obtained by bootstrap estimation to fit regression models using the variables to the left as dependent; categorical variables were coded as having the characteristic = 1, not having = 0.

* Groups based on local traditions for sharing of responsibility for treatment and care of advanced cancer patients.

‡Time from date of histopathological diagnosis to date of inclusion.

3.3.25   Randomized trial of methadone with or without heroin

Van den Brink *et al.* (2003) described a pair of randomized trials, one for 375 inhaling subjects and the other for 174 injecting subjects. Our concern is with the inhaling trial, and its three arms (A = control, B = 12 months of methadone plus heroin, C = six months of methadone followed by six months of methadone plus heroin). This was an open label study, so any restrictions of the randomization would compromise allocation concealment. Neither the precise methods of randomization nor the baseline *p*-values were provided by Van den Brink *et al.* (2003), but their Table 3.9 does list baseline characteristics, by treatment group, but also with these baseline characteristics grouped into categories. We consider the first four groups of baseline characteristics, so physical health is included, but mental health is not. There are then 20 baseline characteristics considered, and three of them had tied values. We discard these three, and consider the other 17, noting that each treatment group would then have, under the null hypothesis, a one in three chance of being the middle value among the three. That is, when the values of a given baseline characteristic are sorted, and then the values replaced by the treatment group to which they correspond, the patterns are ABC, ACB, BAC, BCA, CAB, and CBA.

Of the six possible patterns, each of which should be equally likely, each treatment group is in the middle for exactly two of them. As it turns out, however, A is in the middle for 10 of the 17 comparisons, C is in the middle for five, and B is in the middle for only two. Putting aside the fact that the baseline characteristics need not actually be independent, and that the hypothesis that B tends to be extreme and not in the middle was generated by the data, we proceed with the binomial model. The resulting *p*-value, P{B in the middle in 0, 1, or 2 comparisons}, is artificially low, but this is offset by how much of the information is being ignored, as each comparison could have provided its own *p*-value, except that the authors did not present these. One could compute *p*-values for the binary baseline characteristics, except that even here only proportions, and not counts, are provided, so this would be only approximate. Certainly, one could not compute *p*-values for the continuous measures, such as age.

Table 3.9 Baseline characteristics of 549 heroin addicts who participated in study, according to prescribed treatment reproduced from BMJ, 2003, 327, 310–2, with permission from the BMJ Publishing Group

	Inhaling			Injecting	
	A* (n = 139)	B[†] (n = 117)	C[‡] (n = 119)	A* (n = 98)	B[†] (n = 76)
Age (years)	39.6	40.0	39.1	38.0	39.2
Male (%)	79.1	78.6	81.5	81.6	82.9
Ethnic Dutch (%)	82.7	80.2	80.5	94.9	96.1
Employed (%)	6.5	5.2	12.1	8.2	8.1
Stable housing (%)	90.6	89.7	86.4	84.7	77.6
Regular drug use (years):					
Heroin	16.7	16.9	16.4	15.4	16.6
Methadone	12.4	12.9	11.9	11.7	12.6
Cocaine	8.0	9.3	7.8	10.1	9.6
Amphetamines	1.5	1.4	1.8	3.0	3.1
Drug use in past month (days):					
Herion	25.9	25.9	25.5	25.9	25.2
Methadone	28.7	28.9	29.1	29.1	29.1
Cocaine	15.2	15.2	13.4	18.0	15.5
Amphetamines	0.1	0.1	0.7	1.2	0.9
Previous drug free treatment (%)	59.4	54.7	58.8	67.0	65.8
Ever overdosed (%)	30.9	28.2	29.4	49.0	47.4
Additional need for addiction treatment (%)[§]	66.9	65.8	72.9	63.3	57.9
Physical health:					
Mean MAP-HSS	11.6	10.6	11.8	11.1	12.1
HIV positive (%)	9.9	3.9	5.6	13.3	13.3
Somatic medication (%)	28.8	21.4	24.4	22.5	19.7
Additional need for somatic treatment (%)[§]	29.2	24.8	36.4	39.8	35.5
Mental health:					
Mean SCL-90	70.7	68.4	79.4	72.7	76.3
Ever attempted suicide (%)	17.3	25.6	26.9	40.2	35.5
Psychiatric medication (%)	33.1	32.5	38.7	35.7	34.7

(*Continued*)

Table 3.9   (*Continued*)

	Inhaling			Injecting	
	A*	B†	C‡	A*	B†
	(n = 139)	(n = 117)	(n = 119)	(n = 98)	(n = 76)
Any current DSM-IV diagnosis (%)	27.7	28.2	36.1	34.0	31.6
Additional need for psychiatric treatment (%)§	26.6	26.5	31.9	32.7	39.6
Social functioning:					
Illegal activities in past month (days)	11.2	11.4	8.4	11.5	12.9
Contact with non-users in past month (days)	16.3	15.8	14.1	13.7	12.1
Median No of charges for theft	10.0	6.0	8.0	10.0	15.0
Median time incarcerated (months)	12.0	12.0	10.0	19.0	13.0

The overall baseline p-value is computed as $P\{B$ in the middle 0 times out of $17\} + P\{B$ in the middle 1 time out of $17\} + P\{B$ in the middle 2 times out of $17\}$, which is

$$\left(\tfrac{2}{3}\right)^{17} + 17 \times \tfrac{1}{3} \times \left(\tfrac{2}{3}\right)^{16} + 136 \times \left(\tfrac{1}{3}\right)^{2} \times \left(\tfrac{2}{3}\right)^{15} = 0.044,$$

significant at the usual 0.05 level, despite the fact that so little information from each baseline comparison was used. In conjunction with the unmasked design and the lack of information regarding the specifics of the randomization procedure, it seems that selection bias is a distinct possibility.

3.3.26 Randomized NINDS trial of tissue plasminogen activator for acute ischemic stroke

The National Institute of Neurological Disorders and Stroke (NINDS) conducted a randomized trial of tissue plasminogen activator for acute ischemic stroke (NINDS rt-PA Stroke Study Group, 1995). Mann

(2004) pointed out that in this trial there was a significant imbalance in baseline stroke severity across the treatment groups in one stratum, and that this imbalance favored the active group, possibly to the point of invalidating the findings reported. The FDA (Food and Drug Administration) review (Walton, 1996) reveals that 13 patients were randomized out of order, meaning to an accession number other than the next one available. Of these 13 patients, all received placebo, whereas only two of them should have (the other 11 should have received Activase). In addition, 18 patients were randomized from the wrong stratum, and this changed the assignments of 11 of them. Of these 11, ten should have received Activase but instead received placebo, and only one should have received placebo and instead received Activase. The FDA review (Walton, 1996) states that 'It remains notable however, that of the 22 patients who had treatment changed due to the randomization difficulties that occurred at the treatment sites, 21 of these involved a patient who should have received Activase being changed to receive placebo. Only one of the 22 was changed from placebo to Activase.' Among the patients switched, the proportion of patients switched to placebo from Activase was over 95%.

3.3.27 Norwegian Timolol Trial

Mitchell (1981) critiqued the Norwegian Timolol Trial, and observed that 'Two general points need to be made: firstly, with nearly 2000 randomised patients one would have expected any differences between the groups in common attributes to have been very small; secondly, in such a large trial random variations would have been expected to occur on both sides of equality, so that in some respects the timolol group should have possessed more adverse factors, while for other attributes the placebo group should have been worse off on entry. Neither of these expectations matches the reality: the placebo group was significantly older, had more previous hypertension, more previous diuretic treatment, more heart failure and cardiomegaly, and, above all, more arrhythmias during the index event (ventricular tachycardia or fibrillation, for example, occurred in 14% of the placebo entrants, but in only 10% of the timolol entrants). In a trial of this size this is a very large difference, as are the differences in 'treated hypertension' (22% v 18%) and previous diuretic use (23% v 18%). These sizable differences are significant at the $p = 0.05 - 0.01$ level, so if the groups differed systematically in ways which could influence prognosis one only requires a further one in 10 spin of the wheel of chance to achieve their total mortality figures. While we may have baulked at accepting a one in 1000 possibility that the 54 differential deaths were due to chance, we can more readily conceive of a one in 10 possibility that a type I error has occurred by the operation between two imbalanced groups.'

Some comment is in order here. The p-value for comparing mortality rates (98 in the timolol group and 152 in the placebo group) was 0.001, or one in a thousand. Since some baseline p-values were as low as 0.01 (one in a hundred), Mitchell (1981) divided the one by the other to obtain the one in ten chance mentioned. That is, if there are 1000 lottery tickets, ten winners, and one grand winner, then one can see that P{winning ticket} $= 10/1000 = 0.01$ and P{grand winner} $= 1/1000 = 0.001$, but the conditional probability of holding the grand winning ticket given that your ticket is a winner is P{grand winner | winning ticket} $= 1/10 = 0.1$. So a conditional probability argument appears to have been behind this "one in ten" statement. And yet the situation is even worse than this, because there was more than one baseline imbalance. Consider an analogy with a deck of cards.

Suppose that the ace of spades represents the grand winner, but that every ace is a winner, and every spade is also a winner. Given that you hold an ace, your chances of holding the ace of spades is one in four; and given that you hold a spade, your chances of holding the ace of spades is one in 13. If you hold both, then your chances of holding the ace of spades is 100%, but by selecting only the "predictor" that makes the strongest case, the ace in this case, one would underestimate this and take the chance to be instead one in four. Clearly, then, consideration needs to be extended to all baseline imbalances in the Norwegian Timolol Trial. But there is a more direct way to assess the robustness of the 0.0001 p-value.

Berger (2001) introduced the concept of the depth of statistical significance for this purpose. This depth of statistical significance hinges around the number of patients that would need to switch groups so as to render the p-value no longer statistically significant at whatever alpha level is deemed appropriate. It was shown, in Section 5.1 of Berger (2001), that the depth of statistical significance for a data set with 108 patients in one group, 111 patients in the other group, and a Smirnov test yielding a one-sided p-value of 0.0002 (1/5000) was five. That is, if as few as six patients are strategically selected from each group to switch to the other group, then this one-sided p-value of 0.0002 would become 0.0310 (which exceeds the customary one-sided alpha level of 0.025). Now the basic Norwegian Timolol Trial data set consisted of a 2×2 contingency table, with 152 deaths out of 939 patients in the placebo group, compared to 98 deaths out of 945 patients in the timolol group.

By Fisher's exact test, the one-sided p-value for this data set is 0.0001. If we switch 12 patients from each group to the other group so as to maximally help the placebo group, then we would have 140 deaths in the placebo group and 110 deaths in the timolol group. The p-value is now 0.0214, still significant at the customary one-sided alpha level of 0.025. However, if we switch 13 patients from each group to the other group, then the data set would be 139 deaths in the placebo group and 111 deaths in the timolol group, for a one-sided p-value of 0.0295, which is no longer significant at the customary one-sided alpha level of 0.025. So at this alpha level, the depth of statistical significance is 12.

To put this in perspective, consider that the 24 patients needed to be switched to undo the statistical significance represents only 1.27% of the total 1884 patients. As we will see in Table 4.3.5, as many as half (50%) of the patients can be randomized to the wrong group if selection bias occurs and the block size is two. As low as the initial p-value was, it can easily be explained exclusively by selection bias.

3.3.28 Laparoscopic versus open appendectomy

Hansen *et al.* (1996) reported a randomized trial of laparoscopic versus open appendectomy, and specifically that the rate of complications was significantly lower in the laparoscopic group (2179, 2%) than in the open group (8172, 11%). We will return to this data set to find the depth of statistical significance, but first we mention that Guyatt *et al.* (2002) reported that one of the authors of this 1996 report revealed some aspects of this study that could have led to selection bias. Specifically, the problem occurred when patients were randomized at night, because at night, 'the attending surgeon's presence was required for the laparoscopic procedure but not the open one; and the limited operating room availability made the longer laparoscopic

procedure an annoyance. Reluctant to call in a consultant, and particularly reluctant with specific senior colleagues, the residents sometimes adopted a practical solution. When an eligible patient appeared,
the residents checked the attending staff and the lineup for the operating room and, depending on the personality of the attending surgeon
and the length of the lineup, held the translucent envelopes containing orders up to the light. As soon as they found one that dictated an
open procedure, they opened that envelope. The first eligible patient
in the morning would then be allocated to a laparoscopic appendectomy group according to the passed-over envelope. If patients who
presented at night were sicker than those who presented during the
day, the residents' behavior would bias the results against the open
procedure.'

It is clear that this could lead to the type of selection bias we have been
discussing, so to assess the robustness of the finding one could compute
the depth of statistical significance as we did in Section 3.3.27. But
without the full data, the only data set that can be so manipulated
is binary data, and the only binary data with a claim of significance
is wound infections. There were eight wound infections among the
72 patients in the open group, and only two among the 79 patients
in the laparascopic group. The two-sided p-value by Fisher's exact
test is 0.0482, yet the one-sided p-value is 0.0352, so the depth of
statistical significance is zero. There is no significance at the 0.025
level one-sided.

3.3.29   The Losartan Intervention for Endpoint Reduction in Hypertension (LIFE) Study

Losartan was compared to atenolol in the Losartan Intervention for
Endpoint Reduction in Hypertension (LIFE) Study (Lindholm *et al.*,
2002). Bloom (2002) challenged the validity of the conclusion (which
was that losartan was the more effective of the two) on the basis of
'more Framingham risk scores, more smokers, more atrial fibrillation, worse baseline diabetes control (requiring increased medication),
and more isolated systolic hypertension . . . in patients in the atenolol
group than in the losartan group'. While baseline imbalances do not
by themselves suggest selection bias, it appears that the point being
made by Bloom (2002) was not simply that there are baseline imbalances but rather that these baseline imbalances all went in the

same direction. As will be discussed in Section 6.1, this could suggest selection bias.

The basic data set for the primary composite endpoint (cardiovascular morbidity and mortality) was 103 events among the 586 patients in the losartan group and 139 events among the 609 patients in the atenolol group (Lindholm *et al.*, 2002). The one-sided p-value by Fisher's exact test was 0.0144. Switching one patient from each group to the other group yields a p-value of 0.0205, which is still significant. But switching two patients from each group to the other group yields a p-value of 0.0288, which no longer is significance at the 0.025 level. So the depth of statistical significance is one. Now two patients out of the total of 1195 represents only 0.167% that need to change groups to undo the observed statistical significance, and certainly it is possible that many more than two patients got switched by selection bias.

3.3.30   The Heart Outcomes Prevention Evaluation (HOPE) Study

The Heart Outcomes Prevention Evaluation (HOPE) Study (Sleight *et al.*, 2001) was criticized by Taylor (2002) for similarity in the direction of the baseline imbalances (similar to the criticism in Section 3.3.29). Specifically, Taylor (2002) noted that 'the placebo group contained more individuals with each of the major risk factors: existing peripheral vascular disease ([PVD] 119), previous myocardial infarction (72), stable angina (74), unstable angina (nine), previous cerebrovascular disease (13), left-ventricular hypertrophy (27), raised total cholesteraol (53), and microalbuminuria (52). There were also more men, more smokers, and a longer duration of disease in patients with diabetes.'

The primary endpoint was a composite of myocardial infarction, stroke, and cardiovascular death, and there were 651 events among 4645 patients in the ramipril group compared to 826 events among 4652 patients in the placebo group (HOPE Investigators, 2000). The p-value was less than 0.0001 by either the log-rank test used by the HOPE Investigators (2000) or by the one-sided Fisher's exact test. Switching 51 patients from each group to the other group yields a p-value of 0.0221 by Fisher's exact test (the timing of the events would be needed to compute the log-rank test), and switching 52 patients from each group to the other group would yield a p-value of 0.0253, so the depth of statistical significance (at least when using Fisher's exact test at the level of 0.025 one-sided) is 51. Now 102 patients out of the total of 9291 represents only 1.1% that need to be switched to undo the significance.

3.4   IN SEARCH OF BETTER EVIDENCE

As pointed out by Berger and Weinstein (2004), the overwhelming majority of trials avoid any discussion of selection bias altogether. We are aware of only six trials that reported using the Berger–Exner test (discussed in Chapter 6) and/or one of its variants. Specifically Kroenke *et al.* (2001) stated explicitly that they 'tested for bias in treatment group assignment (using the method described by Berger and Exner) and detected no selection bias'. Van Dijk *et al.* (2002) 'compared 21 baseline characteristics from off-pump patients who had a high likelihood of being randomized to off-pump with the baseline characteristics from on-pump patients with a low likelihood. No significant differences were observed in all the comparisons on the 21 baseline characteristics. Both analyses indicate that there was no selection bias (i.e., the randomization sequence was well concealed).' The ENRICHD Investigators (2003) stated that 'To test for the potential for selection bias that results from research staff being able to predict the next treatment assignment based on unmasking of previous assignments, we used methods developed by Berger and Exner to test for selection bias by examining the association between the predicted probabilities of assignment to the intervention arm (assuming knowledge of sequence of prior allocations) and selected baseline characteristics and event-free survival within each treatment group. All tests were nonsignificant, providing some assurance that any treatment group imbalance on baseline factors and observed treatment effects are not due to selection bias.'

Gerdesmeyer *et al.* (2003) stated that 'The method of Berger and Exner provided strong support against selection bias; comparing baseline CMS [Constant and Murley Scale] values with conditional probabilities that the next treatment is high energy or low energy given knowledge of the sequence of prior allocations within the randomization block, we obtained Pearson correlation coefficients of 0.03 and −0.01, respectively.' The POTS Team (2004) stated that

'Concealment methods followed standard recommendations; no between-treatment group differences at baseline or evidence of statistically identifiable selection biases were apparent. We tested whether there was any selection bias in treatment assignment by examining the probability of each condition within each randomized block (i.e., 0.25 for the first condition in the block and 1.0 for the fourth condition within the block) and tested whether these probabilities interacted with time, treatment condition, and site to predict outcome. No evidence for selection bias was found $(F(3, 281) = 0.06, p = 0.98)$.'

Asarnow *et al.* (2005) stated that 'Screening/enrollment staff were masked to randomization status and sequence and were different from assessment staff. There was also a time delay between screening and randomization (median 21 days). These design features prevented protocol subversion due to selection bias in enrollment that might occur with blocked randomization; we also applied the Berger-Exner test to confirm this expectation.' More elaboration on the results of the testing, and on the delay, might have been helpful, but it certainly appears from the description that the results showed a lack of selection bias. The block size appears to have been two ('To improve balance across conditions in terms of clinician mix and patient sequence, we stratified participants by site and clinician and blocked participants recruited from the same clinicians in pairs according to the time of their enrollment'). This being the case, the delay mentioned above might have been used to ensure that both patients in a block were identified before either one was allocated. This step would certainly be expected to minimize or even eliminate selection bias (Berger and Christophi, 2003). Each of these six trials can therefore be classified definitively as being free of selection bias. There is no need to guess. The rest of the trials, that were not subjected to any meaningful test of selection bias, are generally simply assumed to be free of selection bias, but this would be rather generous. It is probably more reasonable to call these other trials inconclusive regarding the presence of selection bias.

It may be argued that 30 is not a large number of trials, relative to all the trials that have been conducted, so selection bias does not appear to be a major problem. But we have at least reasonable information regarding selection bias for only the 30 trials described in Section 3.3 and the six additional trials discussed in this section, or 36 trials altogether. So the 30 should be compared not to the total number of trials conducted, but rather to the number of trials with any indication, one way or the other, regarding selection bias. The denominator for the 30 should be 30 + 6, or 36. Using this appallingly small number as a denominator, 30 no longer seems so small a numerator, as 30/36 = 83%. Is it true that 83% of all trials are tainted by selection bias? At first glance, this seems highly unlikely, because if this were the case, then one could infer that many of the treatments currently available (even if not 86% of them) are actually useless, or possibly even harmful. Surely such a state of affairs would be readily detected, and if we are not aware that this is the situation, then this cannot be the situation?

One would hope for such transparency, but the reality is that it is unusual for any single physician to have sufficient experience with any given treatment to verify the claims supporting its use. As Penston (2003, page 71) states, 'It is, perhaps, one of the strangest aspects of mega-trials that the supposed benefits of a drug are not observable in routine clinical practice. Given that thousands of patients were required to show any difference, no single clinician would be able to treat a sufficient number of patients in order to detect the difference reported in the studies. Thus, as far as the clinician's experience is concerned, it makes no difference whether or not the result of a mega-trial is valid.' Conspiring with the limited number of patients comprising a clinician's experience with a given treatment is the fact that few, if any, clinicians will treat comparable patients with an alternative treatment to form the basis of a comparison.

Penston (2003, pages 98–99) adds 'Unfortunately, unlike most scientific disciplines in which fraud may be readily detected by attempts at replication, the special circumstances surrounding mega-trials preclude this crucial check on the validity of research. The complexity and opaque nature of large-scale trials, the small treatment differences amenable to manipulation, the enormous potential profits for a successful new drug in common chronic diseases and the whole enterprise under the influence of those with vested interests in the outcome are fertile ground for fraud. But, most crucial of all is the strong likelihood that the fraud will remain undetected. Observations made in routine clinical practice will never be sufficient to disprove the results of a mega-trial, while replication is weakened to such an extent that it no longer has the power to discriminate between authentic and flawed research.' One has to wonder, then, how anybody would know if 83% of the treatments were useless.

Again, this is not the claim being made here, even if the claim were that 83% of the trials are tainted by selection bias. First, even useful and efficacious treatments can be the subject of tainted trials, in which case a true benefit is magnified (as opposed to the illusion of a benefit being manufactured). Second, there is not a one-to-one correspondence between trials and treatments – many treatments are studied with multiple trials. Nevertheless, it would still be a cause for concern if 83% of the trials were flawed in this way alone (to say nothing of the other ways in which trials could be flawed). So is this possible? In fact, this is possible, but this chapter does not provide credible evidence to suggest that this is the case.

As Berger and Weinstein (2004) point out, 'the serendipitous methods by which we encountered trials that we could evaluate for selection bias are not amenable to offering any credible indicator of the true extent to which selection bias occurs in randomized trials, so our examples may represent only the tip of the iceberg'. What this chapter does suggest, however, is that the problem may occur on a regular basis and, to the extent that it results in distortions of trial findings, is deserving of more attention, so that we can better quantify its impact on trial findings, and better minimize this impact. If medical journals required authors of randomized clinical trial reports to disclose sufficient details to allow for an assessment of selection bias, which could be used in determining one aspect of trial quality and rigor, then we would have a much better idea of just how prevalent the problem is. This could be done without violating the HIPAA restrictions (see Geller *et al.*, 2004, for a discussion of these restrictions), because the most basic test for selection bias requires only the patient accession numbers, the restrictions used on the randomization, the treatment allocations, and the response (see Section 6.5).

Note that we have not yet said anything about what impact selection bias has on trial findings. This issue is taken up in Chapter 4, which will complete Part I of the book, which might be taken as a statement of the problem. The remainder of the book will attempt to provide solutions to the problem, in terms of prevention (Chapter 5), detection (Chapter 6), and correction (Chapter 7).

4

Impact of Selection Bias in Randomized Trials

We saw in Chapter 3 that the type of selection bias we described in Chapter 2 appears to have occurred in numerous randomized trials. We did not, however, discuss the impact this selection bias might have in terms of biased parameter estimates, artificially low p-values, inflated Type I error rates, or artificially narrow confidence intervals. All of these problems derive from the induced covariate imbalance, which in turn derives from the prediction of future allocations. Dupin-Spriet *et al.* (2004) quantified the predictability of future allocations in trials using block randomization, whereas Berger (2005a) quantified the resulting covariate imbalance and Berger *et al.* (2003a) quantified the inflation in the Type I error rate associated with selection bias. These issues, specifically prediction of future allocations and the resulting covariate imbalance that occurs when investigators use this knowledge, in conjunction with rough estimates of the potential responses of prospective patients, to make enrollment decisions, will be the focus of this chapter.

4.1 QUANTIFYING THE PREDICTION OF FUTURE ALLOCATIONS: BALANCED BLOCKS

Dupin-Spriet *et al.* (2004) quantified the predictability of future allocations in trials using block randomization, with block sizes ranging

Selection Bias and Covariate Imbalances in Randomized Clinical Trials   V. W. Berger
© 2005 John Wiley & Sons, Ltd.

Table 4.1   Predictability in trials with balanced randomization reproduced from the Drug Information Journal, Vol. 38, p. 129, 2004

Number of arms	Block length	Block composition (*)	Predictability
2	2	1, 1	0.500
2	4	2, 2	0.333
2	6	3, 3	0.250
2	8	4, 4	0.200
2	10	5, 5	0.167
2	12	6, 6	0.143
2	14	7, 7	0.125
2	16	8, 8	0.111
2	18	9, 9	0.100
3	3	1, 1, 1	0.333
3	6	2, 2, 2	0.200
3	9	3, 3, 3	0.143
3	12	4, 4, 4	0.111
3	15	5, 5, 5	0.091
3	18	6, 6, 6	0.077
4	4	1, 1, 1, 1	0.250
4	8	2, 2, 2, 2	0.143
4	12	3, 3, 3, 3	0.100
4	16	4, 4, 4, 4	0.077
4	20	5, 5, 5, 5	0.063

(*) Block composition = number of treatment allocations by treatment arm.

from 2 to 20. Also, both balanced blocks (1:1 allocation) and un-balanced blocks (2:1 allocation) were considered, as was the case in which there are three or four arms, and 1:1:1, 2:2:1, and 1:1:1:1 allocation. The results appear in their Tables 4.1 and 4.2, reproduced here. Berger *et al.* (2003a) distinguished predictable allocations, which are those whose conditional distribution differs from the unconditional distribution specified by the allocation proportions, from deterministic allocations, which are those for which the conditional distribution is degenerate, having a positive probability of only one outcome. That is, a predictable allocation is one for which one can gain an advantage by considering the previous allocations and the restrictions on the randomization, whereas a deterministic allocation can be deduced with certainty given the previous allocations and the restrictions on the randomization.

Table 4.2   Predictability in trials with unbalanced randomization reproduced from the Drug Information Journal, Vol. 38, p. 129, 2004

Number of arms	Block length	Block composition (*)	Predictability
2	3	2, 1	0.444
2	6	4, 2	0.289
2	9	6, 3	0.214
2	12	8, 4	0.170
2	15	10, 5	0.141
2	18	12, 6	0.121
3	5	2, 2, 1	0.240
3	10	4, 4, 2	0.136
3	15	6, 6, 3	0.095
3	20	8, 8, 4	0.073

(*) Block composition = number of treatment allocations by treatment arm.

For example, if there is 1:1 allocation within blocks of size 4, then in the sequence ABAB only the fourth allocation is deterministic, because it is known with certainty that it must be B, given that each block has only two allocations to A and that two allocations to A have already been made in this block. However, the second allocation is predictable, because after observing the first allocation to A, there remain two allocations to B and only one to A. The conditional probability of B, then, is $\frac{2}{3}$, which differs from the unconditional probability of $^1/_2$, as determined by the specified allocation proportions (1:1). The first and third allocations are both unpredictable, because given the prior allocations and the block size of 4, the conditional probabilities are still $^1/_2$ to each of A and B, matching the unconditional probabilities.

The distinction between predictable allocations and deterministic allocations is essential in understanding the results of Dupin-Spriet *et al.* (2004). When tabulating the prediction of future allocations, one could tabulate the frequency of predictable allocations, the frequency of deterministic allocations, or the frequency of correct guesses. This last measure would essentially assign 'partial credit' to predictable allocations in proportion to just how predictable they are. So, for example, consider again the case of 1:1 allocation within blocks of size 4. There are six sequences, {AABB, ABAB, ABBA, BAAB, BABA, BBAA}, each with probability $\frac{1}{6}$. In each of them, the fourth allocation is deterministic. In the first and last sequences, the third allocation is also deterministic. Overall, then, $(6 + 2)/24 = \frac{1}{3}$ of the allocations are deterministic with this allocation scheme. This is the entry tabulated for 1:1 allocation with blocks of size 4 by Dupin-Spriet *et al.* (2004). But certainly there are more predictable allocations than this. In all sequences, the second allocation is predictable, with conditional probability $\frac{2}{3}$ and $\frac{1}{3}$ instead of $\frac{1}{2}$ and $\frac{1}{2}$. So the frequency of predictable allocations is $(6 + 6 + 2)/24 = 7/12$, rather than $4/12$.

The implications of these predictable allocations cannot be ignored, for they increase the success rate in predicting future allocations. With complete ignorance, one would guess right 50% of the time. If one ignored the past allocations except for when the allocation was deterministic, then one would be right $\frac{1}{3} \times \frac{1}{1} + \frac{2}{3} \times \frac{1}{2} = \frac{2}{3}$ of the time. But if one were instead to use the convergent or optimal guessing strategy, guessing that the next allocation will be to the less well-represented group (Matts and Lachin, 1988; Rosenberger and Lachin, 2002, Section 6.5.2), then one would be right $4/12 \times 1/1 + 3/12 \times 2/3 + 5/12 \times 1/2 = 17/24$ of the time, instead of $16/24$ of the time. This quantity, the proportion of predictions that are expected to be correct, may be the most relevant of the three for the purposes of quantifying selection bias. However, the importance of this quantity does not minimize the importance of the other quantities, so we present all three quantities, albeit for a less extensive set of block sizes and compositions than presented by Dupin-Spriet *et al.* (2004).

Some of the calculations are a bit involved, so for the sake of clarity and completeness we tabulate what may be considered worksheets that show the calculations behind the primary table of this section, starting with the simplest case of blocks of size 2, in Table 4.3A. The columns indicate the possible sequences, the number of deterministic allocations (with the deterministic allocations identified in parentheses), the number of predictable allocations (with the predictable

Table 4.3A   Prediction in blocks of size 2 and 1:1 allocation

Block	Deterministic	Predictable	Convergent
AB	1 (no. 2)	1 (no. 2)	$0.5 + 1 = 1.5$
BA	1 (no. 2)	1 (no. 2)	$0.5 + 1 = 1.5$
Total	$2/4 = 50\%$	$2/4 = 50\%$	$3/4 = 75\%$

Table 4.3B   Prediction in blocks of size 4 and 2:2 allocation

Block	Deterministic	Predictable	Convergent
AABB	2 (nos. 3, 4)	3 (nos. 2, 3, 4)	$0.5 + 0 + 1 + 1 = 2.5$
ABAB	1 (no. 4)	2 (nos. 2, 4)	$0.5 + 1 + 0.5 + 1 = 3$
ABBA	1 (no. 4)	2 (nos. 2, 4)	$0.5 + 1 + 0.5 + 1 = 3$
BAAB	1 (no. 4)	2 (nos. 2, 4)	$0.5 + 1 + 0.5 + 1 = 3$
BABA	1 (no. 4)	2 (nos. 2, 4)	$0.5 + 1 + 0.5 + 1 = 3$
BBAA	2 (nos. 3, 4)	3 (nos. 2, 3, 4)	$0.5 + 0 + 1 + 1 = 2.5$
Total	$8/24 = 33\%$	$14/24 = 58\%$	$17/24 = 71\%$

allocations identified in parentheses), and the expected number of correct guesses when the convergent strategy is used, tossing a coin when the allocation is not predictable. We see that half of the allocations are predictable and deterministic when the block size is 2, and that three of every four guesses will be correct. We now consider balanced allocation with blocks of size 4 (2:2 allocation), in Table 4.3B, and note that the proportion of correct guesses always exceeds the proportion of deterministic allocations, but within a block it can be either more than or less than the proportion of predictable allocations. For an AABB block, for example, 75% of the allocations (3/4) are predictable but one of these (the second) would lead to the wrong prediction, so only 62.5% of the allocations would be guessed correctly.

Perhaps the most striking feature when comparing Tables 4.3A and 4.3B is the variation in the magnitude of protection from selection bias when increasing the block size from 2 to 4. When considering only deterministic allocations, as Dupin-Spriet *et al.* (2004) did, there appears to be a large benefit in this increase in block size, as the proportion of deterministic allocations decreases from 50% to 33%. Yet the proportion of predictable allocations actually increases, and the proportion of correct guesses decreases only slightly, from 75% to 71%. In Table 4.3C we consider balanced blocks of size 6. There are 20 ways to select three allocations out of six, and hence 20 possible blocks of this composition, but by symmetry it suffices to present only the 10 that begin with A as the first allocation of the block.

Integrating Tables 4.3A, 4.3B, and 4.3C allows us to construct Table 4.4. The column labeled 'Deterministic' records the proportion

Table 4.3C   Prediction in blocks of size 6 and 3:3 allocation

Block	Deterministic	Predictable	Convergent
AAABBB	3 (nos. 4, 5, 6)	5 (nos. 2–6)	$0.5 + 0 + 0 + 1 + 1 + 1 = 3.5$
AABABB	2 (nos. 5, 6)	5 (nos. 2–6)	$0.5 + 0 + 1 + 0 + 1 + 1 = 3.5$
AABBAB	1 (no. 6)	4 (nos. 2–4, 6)	$0.5 + 0 + 1 + 1 + 0.5 + 1 = 4$
AABBBA	1 (no. 6)	4 (nos. 2–4, 6)	$0.5 + 0 + 1 + 1 + 0.5 + 1 = 4$
ABAABB	2 (nos. 5, 6)	4 (nos. 2, 4–6)	$0.5 + 1 + 0.5 + 0 + 1 + 1 = 4$
ABABAB	1 (no. 6)	3 (nos. 2, 4, 6)	$0.5 + 1 + 0.5 + 1 + 0.5 + 1 = 4.5$
ABABBA	1 (no. 6)	3 (nos. 2, 4, 6)	$0.5 + 1 + 0.5 + 1 + 0.5 + 1 = 4.5$
ABBAAB	1 (no. 6)	3 (nos. 2, 4, 6)	$0.5 + 1 + 0.5 + 1 + 0.5 + 1 = 4.5$
ABBABA	1 (no. 6)	3 (nos. 2, 4, 6)	$0.5 + 1 + 0.5 + 1 + 0.5 + 1 = 4.5$
ABBBAA	2 (nos. 5, 6)	4 (nos. 2, 4–6)	$0.5 + 1 + 0.5 + 0 + 1 + 1 = 4$
Total	$15/60 = 25\%$	$38/60 = 63\%$	$41/60 = 68\%$

of deterministic allocations, and matches the corresponding entries in the tables presented by Dupin-Spriet *et al.* (2004).

4.2   QUANTIFYING PREDICTION OF FUTURE ALLOCATIONS: UNBALANCED BLOCKS

The situation is a little more complicated when the blocks are unbalanced, because in this case there is an apparent contradiction between the convergent guessing strategy and the definition of predictable allocations. Recall that an allocation is predictable if its conditional probability distribution, given the prior allocations and the restrictions on

Table 4.4   Prediction of future allocations with balanced (1:1) blocks

Size	Ratio	Deterministic	Predictable	Correct guesses
2	1:1	50%	50%	75%
4	2:2	33%	58%	71%
6	3:3	25%	63%	68%

the randomization, differs from the unconditional probability distribution. In the case of two treatment arms and 1:1 allocations, this simply means that the conditional probabilities are not 1:1. In this case, the convergent strategy would agree with the definition of a predictable allocation, because it would involve predicting the treatment arm that now has a higher conditional probability, by virtue of being so far less well represented among the previous allocations. But now consider a case in which the block size is 6 and the allocation proportions are 4:2, so that the unconditional probability distribution is $\frac{2}{3}$ to one treatment group, say A, and $\frac{1}{3}$ to the other treatment group, say B.

Now consider the block AABABA. The unconditional and conditional probabilities of A for the six allocations are $(\frac{4}{6}, \frac{4}{6})$, $(\frac{4}{6}, \frac{3}{5})$, $(\frac{4}{6}, \frac{2}{4})$, $(\frac{4}{6}, \frac{2}{3})$, $(\frac{4}{6}, \frac{1}{2})$, $(\frac{4}{6}, \frac{1}{1})$, so clearly the sixth allocation is predictable and the first and fourth allocations are not predictable. But what about the second, third, and fifth allocations? All are strictly predictable by the definition, whereas the third and fifth would not trigger a guess according to the convergent strategy, because the conditional probability is the same for either treatment group. Why, then, are these allocations predictable? The conditional probability of B for this block is never more than 0.5, so if one is going to try to assign a certain type of patient to treatment group B, then one would do so when this conditional probability is at its maximum, 0.5.

That is, one who would attempt to bias the patient selection would do so for the third and fifth allocations by betting that they will be to treatment group B (and selecting patients accordingly) on the basis that the conditional probability of B exceeds the unconditional probability of B. In fact, one might deviate from the convergent strategy and guess B even for the second allocation, despite the fact that its conditional probability is 0.4, which is obviously less than 0.5, because at least it is more than 0.33, the unconditional probability. We will call this guessing strategy the directional strategy, as it is based on the sign (or the direction) of the difference between the conditional and unconditional probabilities. This will differ from the convergent strategy only when the unconditional allocation proportions are different for the treatment groups, or for unbalanced blocks, which we now consider.

First, consider a block of size 3, with 1:2 allocation to A and B (that is, each block has two patients allocated to B and one to A). Then there are three types of block, specifically ABB, BAB, and BBA. We consider these in Table 4.5A. Notice that the second and third

Table 4.5A   Prediction in blocks of size 3 and 1:2 allocation

Block	Deterministic	Predictable	Convergent	Directional
ABB	2 (nos. 2, 3)	2 (nos. 2, 3)	$0 + 1 + 1 = 2$	$0.33 + 1 + 1 = 2.33$
BAB	1 (no. 3)	2 (nos. 2, 3)	$1 + 0.5 + 1 = 2.5$	$0.67 + 1 + 1 = 2.67$
BBA	1 (no. 3)	2 (nos. 2, 3)	$1 + 0.5 + 1 = 2.5$	$0.67 + 0 + 1 = 1.67$
Total	$4/9 = 44\%$	$6/9 = 67\%$	$7/9 = 78\%$	$6.67/9 = 74\%$

allocations are predictable for each block type. This is because whenever the block size is prime, as it is in this case, every allocation after the first is predictable. To understand why, consider that the denominator of the unconditional probabilities is the block size, which is prime, whereas the denominator of the conditional probabilities is the number of allocations remaining in the block. No denominator other than the first can then be a factor of the block size, and so the conditional probability cannot equal the unconditional probability. That is, no integer, when divided by 1 or by 2, can equal $\frac{1}{3}$ or $\frac{2}{3}$. Likewise, no integer, when divided by 4, 3, 2, or 1, can equal $\frac{2}{5}$ or $\frac{3}{5}$, and so with blocks of size 5 and 2:3 allocation to A and B (that is, each block has two patients allocated to A and three to B), every allocation after the first is predictable, as in Table 4.5B.

As expected, the convergent guessing strategy results in more correct guesses than the directional strategy, but not by a terribly large margin. Moreover, after the first allocation is observed, either strategy can be used. If A is observed first, then either strategy would result in 13/16 correct guesses from this point on, whereas if B is observed first, then either would result in 17/24 correct guesses. The only advantage of the convergent strategy, then, is in the prediction of the first allocation. Returning now to the issue of the prime block size, we point out that all allocations after the first are predictable if the allocation proportions are in their lowest terms, even if the block size is not prime. For example, if the block size is 6, and the allocation is 5:1, then every allocation after the first is predictable, because neither $\frac{1}{6}$ nor $\frac{5}{6}$ can be reduced (1 and 6 are relatively prime, as are 5 and 6). On the other hand, $\frac{2}{6}$ and $\frac{4}{6}$ can be reduced, to $\frac{1}{3}$ and $\frac{2}{3}$, respectively, so a block of size 6 with 4:2 allocation could have unpredictable allocations after the third allocation, if the allocation among these first three allocations were 2:1.

Table 4.5B Prediction in blocks of size 5 and 2:3 allocation

Block	Deterministic	Predictable	Convergent	Directional
AABBB	3 (no. 3)	4	$0 + 0 + 1 + 1 + 1 = 3$	$0.4 + 0 + 1 + 1 + 1 = 3.4$
ABABB	2 (nos. 4, 5)	4	$0 + 1 + 0 + 1 + 1 = 3$	$0.4 + 1 + 0 + 1 + 1 = 3.4$
ABBAB	1 (no. 5)	4	$0 + 1 + 1 + 0.5 + 1 = 3.5$	$0.4 + 1 + 1 + 1 + 1 = 4.4$
ABBBA	1 (no. 5)	4	$0 + 1 + 1 + 0.5 + 1 = 3.5$	$0.4 + 1 + 1 + 0 + 1 = 3.4$
BAABB	2 (nos. 4, 5)	4	$1 + 0.5 + 0 + 1 + 1 = 3.5$	$0.6 + 1 + 0 + 1 + 1 = 3.6$
BABAB	1 (no. 5)	4	$1 + 0.5 + 1 + 0.5 + 1 = 4$	$0.6 + 1 + 1 + 1 + 1 = 4.6$
BABBA	1 (no. 5)	4	$1 + 0.5 + 1 + 0.5 + 1 = 4$	$0.6 + 1 + 1 + 0 + 1 = 3.6$
BBAAB	1 (no. 5)	4	$1 + 0.5 + 1 + 0.5 + 1 = 4$	$0.6 + 0 + 1 + 1 + 1 = 3.6$
BBABA	1 (no. 5)	4	$1 + 0.5 + 1 + 0.5 + 1 = 4$	$0.6 + 0 + 1 + 0 + 1 = 2.6$
BBBAA	2 (nos. 4, 5)	4	$1 + 0.5 + 0 + 1 + 1 = 3.5$	$0.6 + 0 + 0 + 1 + 1 = 2.6$
Total	$15/50 = 30\%$	$40/50 = 80\%$	$36/50 = 72\%$	$35.2/50 = 70.4\%$

Table 4.6   Prediction of future allocations with two arms

Size	Ratio	Deterministic	Predictable	Convergent	Directional
2	1:1	50%	50%	75%	75%
3	1:2	44%	67%	78%	74%
4	2:2	33%	58%	71%	71%
5	2:3	30%	80%	72%	70.4%
6	3:3	25%	63%	68%	68%

We can use the results of Tables 4.5A and 4.5B to add additional entries to Table 4.4. We do so in Table 4.6, and notice that there are many more predictable allocations when the block is unbalanced than when it is balanced. Despite this, the proportion of correct guesses is not dramatically different whether the blocks are balanced or unbalanced. This is probably good news, because it means that not too much is lost when the blocks are unbalanced. However, the proportion of correct guesses is not overly sensitive to the block size either, and this is probably bad news, because it means that increasing the block size will not by itself do too much to minimize the problem of selection bias.

4.3   QUANTIFYING COVARIATE IMBALANCE RESULTING FROM SELECTION BIAS

In Sections 4.1 and 4.2 we studied the prediction of future allocations when randomized blocks are used with equal or unequal allocation. This prediction of future allocations is a problem only to the extent that it leads to a separation in the covariate distributions across the treatment groups. In this section we quantify the covariate imbalance that would result from this prediction of future allocations, depending on the extent to which an investigator exploits this prediction. Note that 'even small imbalances in important prognostic factors could overwhelm treatment differences, either producing apparent treatment effects when none in fact are present or masking true treatment differences when they do exist' (Green and Byar, 1984).

Like Berger (2005a), we define G be the value of the conditional probability required to bias patient selection. To clarify this definition further, consider a binary covariate X, with values 1 for strong patients (likely to respond to either treatment) and 0 for weak patients (unlikely to respond to either treatment). If $P\{E\} > G$, then a strong patient is enrolled; if $P\{E\} < 1 - G$, then a weak patient is enrolled; and if $1 - G < P\{E\} < G$, then the first eligible patient who consents is enrolled. A consequence of this definition is that it does not apply only when there is selection bias, because the case of no attempt to bias the enrollment, even in the presence of prediction of future allocations, is represented by $G = 1.00$. Also, if G is less than 1.00, but close enough to 1.00 that it exceeds every conditional probability less than 1.00 (there are only so many values that these conditional probabilities can assume), then the investigator attempts to bias the enrollment only when he or she is certain of the next allocation. In a very real sense, then, G determines the propensity of a given investigator to engage in subversion of the randomization, and can be calibrated to correspond to what Feigenbaum and Levy (1996) called 'saints' ($G = 1.00$), 'jerks' (G just under 1.00, say $G = 0.99$), and 'careful crooks' ($G = 0.50$).

The actual covariate imbalance induced by prediction of future allocations depends on both the extent to which these future allocations are predictable and the value of G. Phrased differently, the extent to which the future allocations are predictable represents the potential for selection bias, and places a limit on how unbalanced covariates can be by this mechanism. Then G represents the extent to which the potential is realized. Suppose that in the population from which patients are sampled $X = 1$ in $100P\%$ of the patients, and $X = 0$ in the other $100(1 - P)\%$. In the absence of selection bias, either because $G = 1$ or because there is no prediction of future allocations, we would expect this distribution of X overall to hold roughly within each treatment group. The effect of any selection bias that may occur would be to increase the proportion of $X = 1$ patients in one arm, say the active arm, and to decrease the proportion of $X = 1$ patients in the other arm, say the control arm. We define the imbalance of covariate X to be the difference, across groups, in the proportion of $X = 1$ patients. Now if $G < 1$, then the patient selection, be it $X = 0$, $X = 1$, or the first qualified patient, will depend on the value of the conditional probability of allocation to the active (experimental) treatment group. Of course, this conditional probability, which we call $P\{E\}$, is itself unbalanced across the treatment groups, as more often than not the higher values will be assigned to the active treatment group and the lower values will be assigned to the control group.

Berger (2005a) provides intricate calculations of the covariate imbalance as a function of the block size and G; which we reproduce here. When the block size is 2, P{E} can assume only the values 0 (the second allocation in an EC block), 0.5 (the first allocation in a block), and 1 (the second allocation in a CE block). This means that the only critical threshold for G is 1.00. Suppose that these are N Blocks (2N patients Total), if $G = 1.00$, then there is no selection bias and there is no covariate imbalance. That is, the (expected) imbalance is 0, because the number $X = 1$ patients is each treatment group has a binomial distribution with parameters N and P. If, however, $G < 1$, then regardless of how much less than 1.00 it is, the second allocation in each block will be used to select $X = 1$ patients (P{E} = 1 for a CE block) or $X = 0$ patients (P{E} = 0 for an EC block).

Suppose that there are k CE blocks, and $N - k$ EC blocks. Then the active group will have k patients from CE blocks (all of which will have an X value of 1) and $N - k$ patients from EC blocks. The number of $X = 1$ patients in the E group is then $k + B(N - k, p)$, where $B(N - k, p)$ is a binomial random variable with parameters $N - k$ and p. In contrast, the C group will have $N - k$ patients from EC blocks (all $X = 0$), and k patients from CE blocks, so the number of $X = 1$ patients in the C group is B(k,p). The expected imbalance given the value of k is then:

$$P\{X = 1|E, k\} - P\{X = 1|C, k\} = E[(k + B(N - k, p))/N]$$
$$-E[B(k, p)/N]$$
$$= (k + Np - kp - kp)/N$$
$$= [k + p(N - 2k)]/N.$$

This expected imbalance ranges from p when $k = 0$ to $1 - p$ when $k = N$, but perhaps what is most relevant is that when k assumes its expected value of N/2 this expected imbalance is $1/2$, or 50%. That is, there would be, on average, 50% more $X = 1$ patients in the E group than in the C group, perhaps 70% vs. 20% or 80% vs. 30%. This expected imbalance of 50% is specific to the situation of blocks of size two and $G < 1$. If the block size is larger than two, then the computations are more involved. With blocks of size four and 2:2 allocation, for example, the possible values of P{E} are 0.00, 0.33, 0.50, 0.67, and 1.00. This means that the relevant thresholds are 0.67 and 1.00.

Table 4.7 Joint distribution of the block position (BP), ranging from one to the size of the block, and the probability P{E} of allocation to the experimental group E, overall (all 24 patients) and by group (12 patients in each of the E and C groups), six blocks of size four (one of each type: EECC, ECEC, ECCE, CEEC, CECE, CCEE) reproduced by permission of Wiley-VCH

BP	P{E}	P{C}	Configurations (number)	Total (24)	E Group (12)	C Group (12)
1	2/4	2/4	all six (6)	6	3	3
2	1/3	2/3	EECC, ECEC, ECCE (3)	3	1	2
2	2/3	1/3	CCEE, CECE, CEEC (3)	3	2	1
3	0/2	2/2	EECC (1)	1	0	1
3	1/2	1/2	ECEC, ECCE, CECE, CEEC (4)	4	2	2
3	2/2	0/2	CCEE (1)	1	1	0
4	0/1	1/1	EECC, ECEC, CEEC (3)	3	0	3
4	1/1	0/1	CCEE, CECE, ECCE (3)	3	3	0

Table 1 of Berger (2005), reproduced as Table 4.7 here, presents the joint distribution of the block position within each block and P{E}, both overall and within each treatment group, when the block size is four and there is 2:2 allocation within each block. In this case, the block position (BP) ranges from one to four, and P{E} depends on both BP and the sequence within the block, which can be any of the six types, {CCEE, CECE, CEEC, ECCE, ECEC, EECC}. Also presented is P{C}, which is $1 - P\{E\}$ for a two-arm trial, as is assumed in Table 4.7. One can readily see that as expected, the E group will have a disproportionate number of patients randomized when P{E} is large, and the C group will have a disproportionate number of patients randomized when P{E} is small and P{C} is large.

Table 2 of Berger (2005), reproduced as Table 4.8 here, presents the marginal distribution of P{E} within each treatment group, when the block size is four. In addition, different values of G are considered, to study the consequences of different biasing strategies. The P{E} imbalance across treatment groups is retained from Table 4.7, but stands out more now, as we have 'integrated out' the BP. When $G = 1$, there is no attempt to bias the selection, and so for any value of P{E} patients will be selected as they come, meaning that the probability of a 'good responder' or a strong patient ($X = 1$) is the population probability P. When G is between $2/3$ and 0.99, however, there is

Table 4.8   Marginal distribution of the probability P{E} of allocation to
the experimental group E by group, six blocks of size four (one of each type:
EECC, ECEC, ECCE, CEEC, CECE, CCEE). G is the P{E} value above which
'good responders' will be selected with probability one, and with probability
zero for P{E}<1-G

P{E}	P{C}	E Group	C Group	G = 1.00	0.67 < G < 0.99	0.5 < G < 0.66
0.00	1.00	0/12	4/12	P	0	0
0.33	0.67	1/12	2/12	P	P	0
0.50	0.50	5/12	5/12	P	P	P
0.67	0.33	2/12	1/12	P	P	1
1.00	0.00	4/12	0/12	P	1	1

selection bias, because when P{E} = 0 a weak patient ($X = 0$) will
be selected with certainty. When G is between $1/2$ and 2/3, even the
allocations for which P{E} = 1/3 or 2/3 will be biased in this way, to
create an advantage for the E group. For no value of G will there be a
bias when P{E} = 0.5.

With blocks of size six and 3:3 allocation, the possible values of
P{E} are 0.00, 0.25, 0.33, 0.40, 0.50, 0.60, 0.67, 0.75, and 1.00.
This means that the relevant thresholds are 0.60, 0.67 and 0.75, and
1.00. Table 3 of Berger (2005), reproduced here as Table 4.9, is of
the same structure as Table 4.7, except that it treats the case of 3:3
allocation within each block of size six. For this same case, Table 4 of
Berger (2005), reproduced here as Table 4.10, is of the same structure
as Table 4.8.

Table 5 of Berger (2005), reproduced here as Table 4.11, summa-
rizes the results from the cases considered (balanced blocks of size
two, four, and six). As mentioned, the covariate imbalance induced by
selection bias for balanced blocks of size two is 50% for any value of
G below 1. With balanced blocks of size four, however, the covariate
imbalance induced by selection bias depends on G. Specifically, the
covariate imbalance is 0% for $G = 1.0$, 33% for $0.67 < G < 1.00$,
and 42% for $0.5 < G < 0.67$. Likewise, with balanced blocks of size
six, the covariate imbalance induced by selection bias depends on
G. Specifically, the covariate imbalance is 0% for $G = 1.0$, 29% for
$0.75 < G < 1.00$, 33% for $0.67 < G < 0.75$, 38% for $0.60 < G < 0.67$, and 41% for $0.50 < G < 0.60$. For values of G close to 0.50,
the increase in block size does not result in very much reduction in the
covariate imbalance. This is a concern, because while the block size is

Table 4.9 Joint distribution of the block position (BP), ranging from one to the size of the block, and the probability P{E} of allocation to the experimental group E, overall (all 480 patients) and by group (240 patients in each of the E and C groups), 80 blocks of size six (four of each of the 20 configurations: EEECCC, EECECC, ...) reproduced by permission of Wiley-VCH

BP	P{E}	P{C}	Total (480)	E Group (240)	C Group (240)
1	3/6	3/6	80	40	40
2	2/5	3/5	40	16	24
2	3/5	2/5	40	24	16
3	1/4	3/4	20	5	15
3	2/4	2/4	40	20	20
3	3/4	1/4	20	15	5
4	0/3	3/3	10	0	10
4	1/3	2/3	30	10	20
4	2/3	1/3	30	20	10
4	3/3	0/3	10	10	0
5	0/2	2/2	20	0	20
5	1/2	1/2	40	20	20
5	2/2	0/2	20	20	0
6	0/1	1/1	40	0	40
6	1/1	0/1	40	40	0

Table 4.10 Marginal distribution of the probability P{E} of allocation to the experimental group E by group, 80 blocks of size six (20 of each type). G is the P{E} value above which "good responders" will be selected with probability one, and with probability zero for P{E} < 1-G. For 1-G < P{E} < G, this probability P matches the population ratio reproduced by permission of Wiley-VCH

P{E}	P{C}	E Group	C Group	G = 1.00	G = 0.99	G = 0.70	G = 0.65	G = 0.50
0.00	1.00	0/240	70/240	P	0	0	0	0
0.25	0.75	5/240	15/240	P	P	0	0	0
0.33	0.67	10/240	20/240	P	P	P	0	0
0.40	0.60	16/240	24/240	P	P	P	P	0
0.50	0.50	80/240	80/240	P	P	P	P	P
0.60	0.40	24/240	16/240	P	P	P	P	1
0.67	0.33	20/240	10/240	P	P	P	1	1
0.75	0.25	15/240	5/240	P	P	1	1	1
1.00	0.00	70/240	0/240	P	1	1	1	1

Table 4.11   Expected covariate imbalance as a function of the block size and G. G is the P{E} value above which "good responders" (covariate = 1) will be selected with probability one, and with probability zero for P{E} < 1-G. For 1-G < P{E} < G, this probability P of selecting "good responders" (covariate = 1) matches the population ratio reproduced by permission of Wiley-VCH

Block Size	0.50<G <0.60	0.60<G <0.66	0.67<G <0.75	0.75<G <0.99	G=1.00
2	50%	50%	50%	50%	0%
4	42%	42%	33%	33%	0%
6	41%	38%	33%	29%	0%

at the discretion of the sponsors of the trial, the value of G obviously is not (other than through the choices the sponsors make concerning which investigators conduct the trial; see Section 5.2).

4.4   QUANTIFYING THE BIAS RESULTING FROM COVARIATE IMBALANCE

The covariate imbalance discussed in Section 4.3 is not itself a problem except to the extent that it leads to biased parameter estimation, inflated Type I error rates, overly narrow confidence intervals, or overly optimistic posterior probabilities. The extent to which these problems occur depends on more than just the extent to which the covariate is unbalanced; one needs to know also how predictive the covariate is for the response. At one extreme, suppose that the selection covariate (that is used as the basis of defining 'strong' and 'weak' patients) turns out to be unrelated to all other covariates and also to the outcome itself. In this case, the only covariate that is unbalanced happens to be a bad covariate, in the sense of not predicting outcomes at all. If this covariate is unbalanced across treatment groups because future allocations were predictable, then selection bias has occurred, but this selection bias that has occurred did not lead to biased parameter estimation, inflated Type I error rates, overly narrow confidence intervals, or overly optimistic posterior probabilities.

At the other extreme, if the covariate is a perfect predictor of ultimate response, then its imbalance directly leads to the problems discussed

Table 4.12 Probability of a significant result, one investigator reproduced from Stastica Sinica 4, 1994, p 222

	$\alpha = .05$			$\alpha = .01$		
$\gamma = \eta/\sigma$	$2k = 50$	$2k = 100$	$2k = \infty$	$2k = 50$	$2k = 100$	$2k = \infty$
.10	.063	.063	.065	.013	.014	.014
.20	.079	.080	.083	.018	.019	.020
.30	.098	.100	.107	.024	.025	.028
.40	.121	.124	.134	.033	.034	.039
.50	.147	.152	.166	.044	.046	.053

above. For example, the Type I error rate could be made arbitrarily close to 1.00 by increasing the sample size and ensuring that the difference in response rates across treatment groups becomes statistically significant. Proschan (1994) studied the Type I error inflation associated with using one large block, as a function of sample size and how strong (or weak) a patient can be selected for inclusion in a preferred treatment group (in other words, how well the selection covariate predicts the outcome variable). The results appear in Table 4.12 of Proschan (1994), reproduced here.

For covariates that are neither useless not perfect predictors, the results are somewhere in between these two extremes. One set of scenarios was studied by Berger *et al.* (2003a). Specifically, the blocks were all balanced, and ranged from size 4 (2:2 allocation) to 8 (4:4 allocation), G was 0.50 or 0.99, and the strength of the covariate varied as follows. The (continuous, normally distributed) responses of weak, average, and strong patients was taken to be $N(-D, 1)$, $N(0, 1)$, and $N(D, 1)$, so D corresponds to η of Proschan (1994), measures the strength of the covariate, and ranges from 0.0 to 2.0 (twice the variance) by 0.5. The procedures studied were randomized blocks of fixed size, varied block sizes, and the maximal procedure (Section 5.3.4). As expected, when $D = 0.0$ the Type I error rate was the nominal rate, 0.05, regardless of other parameter values. But when $D > 0.0$, the inflation in the Type I error rate was rapid. For example, even if $G = 0.99$ (the situation likely to result in the least selection bias subject to $G < 1.00$), the fixed block size procedure had Type I error rates of 0.05, 0.21. 0.50, 0.77, and 0.93 for $D = 0.0, 0.5, 1.0, 1.5$, and 2.0, respectively. This inflation of the Type I error rate was reduced as the block size increased.

Compared to the randomized block procedure with fixed block sizes, the maximal procedure was more resistant to inflation of the Type I error rate in every case studied. Sometimes, the difference was quite pronounced. For example, if $G = 0.99$, the largest imbalance allowed is 4 (corresponding to blocks of size 8), and $D = 1.5$, then a nominal 0.05 test would have an actual Type I error rate of 0.19 for the maximal procedure and 0.43, more than twice as much, for the randomized blocks procedure. Perhaps surprisingly, varying the block size did not always result in a better procedure than using fixed block sizes, but the line of demarcation between the situations more favorable to each is clearly drawn. Specifically, when $G = 0.05$ the fixed block size were better for any block size and for any value of D, whereas for $G = 0.99$ the varied block sizes were better for any block size and for any value of D. As Berger *et al.* (2003a) explained, this is because the convergent strategy tends to work better, not worse, when the block sizes are varied, than when they are held fixed. See also Rosenberger and Lachin (2002) for additional explanation of this phenomenon.

Ivanova *et al.* (2005) also studied the magnitude of alpha inflation caused by selection bias. Specifically, in their Table III, based on a simulation study, they found that the true Type I error rate could be $0.54, 0.29, 0.62$, or 0.34, depending on how inclined the investigator is to bias patient selection, the common response rate in each treatment group, and the amount by which better responders respond better than poor responders. The cases studied include a common response rate of 50% or 70% and a cutoff for G of 0.50 or 0.99. In each case, there was a 20% better response rate for good responders than for medium responders and a 20% better response rate for medium responders than for poor responders. That is, the common response rates (independent of treatment group) were 30% for poor responders and 70% for good responders (when medium responders had a 50% response rate) or 50% for poor responders and 90% for good responders (when medium responders had a 70% response rate). For a 50% response rate for medium responders, the true alpha level of a nominal 0.05-level test was 29% when $G = 0.99$ and 54% when $G = 0.50$. For a 70% response rate for medium responders, the true alpha level of a nominal 0.05-level test was 34% when $G = 0.99$ and 62% when $G = 0.50$. These simulations were based on 5000 runs, and each pseudo-sample had a sample size of 192. Clearly, a larger sample size would exacerbate, and not solve, this problem.

Part II
Actions to be Taken to Improve the Reliability of Medical Studies

Preventing Selection Bias in Randomized Trials

We saw in Chapter 2 that the type of selection bias we consider requires certain key conditions to exist. Each of these conditions represents a potential opportunity to intervene and prevent selection bias from occurring. Specifically, as we have seen, selection bias occurs when investigators are able to predict upcoming treatment assignments, and then use this knowledge, in conjunction with rough estimates of the potential responses of prospective patients, to make enrollment decisions. Expected responders, for example, may be enrolled or denied enrollment as the active treatment is or is not deemed likely to be allocated next (Blackwell and Hodges, 1957). Selection bias induces confounding to the extent that there is a resulting separation of the distribution of the potential responses across the treatment groups. Therefore, selection bias cannot occur if either there is no advance knowledge of upcoming allocations or the investigator is unwilling or unable to use this advance knowledge to make enrollment decisions. Likewise, selection bias does not create confounding, even if it occurs, if it does not result in a separation of the distribution of the potential responses across the treatment groups.

Our goal, then, is to simultaneously minimize advance knowledge of upcoming treatment assignments, minimize the use of whatever advance knowledge we could not prevent to make enrollment decisions, and minimize the impact on baseline balance across treatment groups of any such biased enrollment decisions. Working backwards, we address the last issue first.

Selection Bias and Covariate Imbalances in Randomized Clinical Trials   V. W. Berger
© 2005 John Wiley & Sons, Ltd.

5.1 MINIMIZING THE IMPACT OF SELECTION BIAS

The potential impact of selection bias depends on the heterogeneity, with respect to potential responses, of the study population. At one extreme, consider a study performed on clones who were not only born identical to each other, but in fact had identical experiences after birth as well. In such a case, all treatment groups would be identical even if attempts to bias the allocation were based on complete knowledge of the upcoming treatments. There simply is no opportunity to exploit differences in patient characteristics to create different treatment group profiles if there are no differences in the patient characteristics. In other words, we need not worry about selection bias in such a case, even though it may well occur, because it would have no impact on the trial. The only reason to even consider selection bias in such a situation would be to assess its potential for occurring, which would be of interest only for future studies not performed in clones.

At the other extreme, each patient would be either a non-responder who would respond to no treatment under study or a responder who would respond to any treatment under study (Berger *et al.*, 2003b), and so any appearance of a treatment effect would be an illusion created by the fact that more responders were funneled to one group than to another. This illusion is facilitated by selection bias, of course, and the point is that selection bias in this situation would be likely to have a profound effect on the reliability of the results. We see that the more homogeneous the study group, at least within strata, the smaller the potential impact of selection bias can be.

5.2 BIASED SELECTION OF INVESTIGATORS

We now address preventing selection bias directly, rather than trying to minimize its impact. One step to take involves, of all things, selection bias applied to the selection of investigators. Specifically, it seems reasonable to check the track record of selection bias for each investigator who is considered for conducting a clinical research study. It would not be sufficient, of course, to equate all investigators on the basis of being unaware of any accusations of selection bias. Rather, one would need to be more proactive, and evaluate the extent to which their previous trials appeared suspicious for selection bias. To be clear, it is the propensity to engage in biased patient selection, rather than the impact of any such selection bias, that needs to be considered when selecting investigators. One would then bias the selection of investigators towards the inclusion of those investigators who do not appear to bias the selection of their studies in the ways discussed previously. Presently, this method for selecting investigators does not appear to be in use, nor could it be without the availability of pertinent data with which to make such decisions. This suggestion, then, actually represents a multitude of suggestions, including that such data be made available by trial sponsors and that such data, once made available (or conditional on their being made available), be used by other sponsors.

5.3 MINIMIZING THE PREDICTION OF FUTURE ALLOCATIONS

Having made the study population as homogeneous as possible (at least within strata), and having strategically selected investigators, we are now in a position to limit the potential for prediction of future treatment allocations. True masking and allocation concealment would, by their very definitions, accomplish the goal of a complete elimination of any such prediction. Moreover, true masking would be sufficient to ensure true allocation concealment.

However, as we saw in Chapter 2, claims of masking indicate only that steps were taken in an effort to ensure masking, and claims of allocation concealment indicate only that steps were taken in an effort to ensure allocation concealment. In neither case can one be certain that these steps were successful. Even in 'masked' trials, meaning trials planned as masked or for which masking is claimed, true masking is rarely attained (Day, 1998; Berger and Exner, 1999). This is why Senn (1995) warned 'that investigators should delude neither themselves, nor those who read their results, into believing [...] simply because some aspects of their trial were double-blind that *therefore* all the virtues of such trials apply to all their conclusions'. For our purposes, the distinction between a masked study and an unmasked study is that in the former case only some of the prior allocations will be known at the time of the next allocation.

One threat to masking in trials planned as masked is the intentional unmasking of the treatment assignment for a patient who requires emergency rescue medication for a serious adverse event. Sometimes the nature of the rescue medication will depend on the study medication, and so for ethical reasons the masking needs to be broken (for this patient). Because of this, it has become customary to break the blind for any patient requiring rescue medication even if the nature of the rescue medication does not, in fact, depend on the study medication (Ayala and MacKillop, 2001). This represents an unnecessary threat to masking, and should probably be avoided to the extent possible. Another threat to masking in trials planned as masked is forcing the randomization to be deterministic, possibly because a given center is out of stock for all treatments but one (McEntegart, 2003). Minimizing such unnecessary unmasking for either reason (intentionally breaking the masking or running out of stock) would help to keep masked trials as well masked as possible.

Also, 99 out of 159 trials published in the *British Medical Journal* between January 1997 and June 2001 (62%) were planned as unmasked (Kjaergard and Als-Nielsen, 2002). So while it is certainly helpful to mask to the extent possible, there is also a need to minimize the extent to which knowledge of past allocations (the lack of masking) leads to fruitful prediction of future allocations. That is, we need to minimize the patterns present in the allocation sequence. One could use unrestricted randomization (Schulz and Grimes, 2002b) to eliminate all patterns and all prediction. One concern with unrestricted randomization is the possibility that at some point in time the treatment group sizes will differ enough that the time and treatment effects are confounded. Hallstrom and Davis (1988) point out that

It is preferable to maintain these equal groups over time. If the enrollment period is long enough that changes in potential confounding variables such as new ancillary treatments or new referral criteria can occur, maintaining the equal groups during the course of the study reduces a potential source of bias. In trials in which the patients enter sequentially, a modification of strict random assignment is used to ensure that the groups will have approximately equal size and will be balanced over time.

Indeed, there have been trials which had patient characteristics change during the recruitment process (Gansky and Koch, 2001). Preventing this from causing chronological bias (Matts and McHugh, 1983), or confounding of the time and treatment effects, requires restrictions on the randomization.

5.3.1 The trade-off between selection bias and chronological bias

To prevent treatment effects from being confounded with time effects, some sort of restriction on the randomization would be required. For example, one may specify that the final group sizes are the same (equal allocation), or that the group sizes never differ by more than a given amount. In practice, most trials use some sort of restricted randomization to force this type of balance. If enrollment is sequential over time, then an unintended consequence of these restrictions is that they create precisely the patterns that, along with unmasking, can lead to prediction of future allocations, which in turn can lead to selection bias. In fact, the more restrictive the allocation procedure, the greater is the potential for selection bias. This is because each restriction represents the potential to predict a future allocation from a previous one, or a set of previous allocations. This is easiest to see with the randomized block design with a given block size, because the set of restrictions on the randomization can, for this design, be summarized with a single number, the common block size. Specifically, the randomized block procedure is characterized by its forced returns to perfect balance at various points during the trial.

Specifically, there must be an equal number of patients allocated to each treatment group within each block. Technically, this is not true, as sometimes allocation is not 1:1, but even here whatever proportions are prescribed must be adhered to within each block. But when there is 1:1 allocation, the equality of numbers allocated within each block translates to equality overall at the completion of each block. This is demonstrated nicely, for the case of a block size of four, in Table 2 of Beller, Gebski, and Keech (2002), reproduced as 5.1.

What is shown is the generation of the allocation sequence by means of randomized blocks of size four. There are six types of block of size four, so one could associate each of these with a number from one to six, as shown in the second column, and then randomly select numbers from one to six. While there are more efficient ways of doing this, what is shown in the first column is selection by randomly generating a number from 0 to 9, and then ignoring the numbers other than 1, 2, 3, 4, 5, and 6. In the table, the 8 is ignored, because it does not correspond to any specific type of block of size four. Such replacement randomization was also discussed by Berger *et al.* (2003a) in the

Table 5.1   The permuted block method of randomisation for a block size of four, with A and B being treatment groups (A = intervention and B = control, for example) Beller, E. M., Gebski, V. and Keech, A. C. 'Randomization in clinical trials'. MJA 2002; 177:566–8. Copyright 2002. *The Medical Journal of Australia* reproduced with permission

Random number sequence	Permuted blocks	Randomisation list
1	1. AABB	1 { A A B B
4	2. ABAB	
8	3. ABBA	
6	4. BBAA	4 { B B A A
5	5. BABA	
(etc)	6. BAAB	6 { B A A B
		5 { B A B A

A random number sequence is generated from a statistical textbook or computer. Each possible permuted block is assigned a number (1 to 6 in the above example). Using each number in the random number sequence in turn selects the next block, determining the next four participant allocations. Numbers in the random number sequence greater than the number of permuted block combinations (7, 8, 9 and 0 in the above example) are not used to select blocks.

context of generating the maximal procedure, which will be discussed in Section 5.3.4.

When using randomized blocks with 1:1 allocation to two treatment groups, the imbalance in the size of the treatment groups will never exceed half the block size (Frane, 1998), which is often twice the number of treatment arms (Pocock and Lagakos, 1982). That is, in the common case of a two-arm trial, it would be common to use blocks of size 4 (as in the table above), so that within each block two patients are allocated to each treatment group. Likewise, with three treatment arms there would often be blocks of size 6.

But basing the block size on the number of treatment arms conceals the true rationale for the strategic selection of a block size. The larger the block size, the less restrictive the randomization, and therefore the less predictable are the future allocations given knowledge of the past ones. However, there is also the downside of a greater possibility of gross imbalance at some point during the trial between the numbers of patients allocated to each treatment group. In fact, the imbalance can be as large as half the block size. We see, then, that smaller block sizes result in greater balance over the course of the trial, but at the expense of more prediction. One solution to the problem of optimal selection of block sizes is to vary the block sizes. We will study this in Section 5.3.3, but first we need to define some notation.

5.3.2   Notation

We use the same notation as Berger *et al.* (2003a). Specifically, we consider a trial with two treatment groups, say A and B. Let D be the allocation sequence (of length $2N$, the number of patients allocated), with $X_i(D) = 1$ or 0 as D assigns the ith patient to treatment arm A or B (this is a change from the notation of Section 4.3, when X was used to denote a covariate). We note that the randomization procedure, be it randomized blocks, varying blocks, or some other procedure, is a probability distribution used to select D. After k patients have been enrolled, D has allocated $S_{k,A}(D) = \sum_{i=1}^{k} X_i(D)$ and $S_{k,B}(D) = k - S_{k,A}(D)$ patients to treatment arms A and B, respectively. The numerical imbalance is then $I_k(D) = S_{k,A}(D) - S_{k,B}(D)$. The 'position' or 'location' $\{k, I_k(D)\}$ of the randomization, be it for the entire trial or for just the stratum, is a random walk on the Cartesian plane, starting at (0,0), moving right if $X_i(D) = 1$ or up if $X_i(D) = 0$, and ending at $\{2N, I_{2N}(D)\}$. This random walk is observed as it unfolds for an unmasked trial, but is concealed for a masked trial.

An allocation procedure, say P, consists of a set $\Gamma(P)$ of $n(P)$ sequences, and a probability distribution, often discrete uniform, on $\Gamma(P)$. Clearly, the extent of chronological bias depends on both D and P, but it also depends on time trends that occur during the study. Nevertheless, we consider only the contribution of D and P to chronological bias. These may be measured by the largest imbalance over the course of the trial, $I(D) = \max_{1 \le k \le 2N} |I_k(D)|$ and

$I(P) = \max_D \in \Gamma(P)I(D)$. **Condition T** specifies terminal balance, $I_{2N}(D) = 0$, or equal allocation to the two treatment groups. The extent of chronological bias to be allowed is measured by the maximum tolerated imbalance (MTI), which must be chosen to balance the need to minimize the prediction of future allocations based on past ones against the need to minimize imbalance and the corresponding chronological bias (Senn, 2000; Atkinson, 2001). **Condition B** specifies adherence to the MTI of b throughout the trial, $|I_k(D)| \le b$ for $k = 0, 1, 2, \ldots, 2N$.

Allocation procedure P will often be made explicit in the study protocol, and we consider it, but not the realized allocation sequence D, to be known to the investigator even before the recruitment of patients begins. If P is uniform, then by definition each $D \in \Gamma(P)$ has probability $1/n(P)$. In this case, patient i is allocated to treatment A with probability $P_{i,P}\{A\} = \sum_D \in \Gamma(P)X_i(D)/n(P)$. Given the path $\{1, I_1(D)\}, \{2, I_2(D)\}, \ldots, \{i - 1, I_{i-1}(D)\}$ of random walk D, or prior allocations, this probability becomes $P_{i,P}\{A|I_{i-1}(D)\} = n_P(i, I_i)/n_P(i - 1, I_{i-1})$, where $n_P(k, I_k)$ is the number of paths from (k, I_k) to $(2N, 0)$. This posterior probability has been derived for the randomized blocks procedure to detect selection bias (Berger and Exner, 1999). Intuitively, the ith allocation of P is deterministic if it is determined by $\Gamma(P)$ and the previous allocations; that is, if $X_i(D)$ is constant for all $D \in \Gamma(P)$ that match the observed initial segment (of length $i - 1$) of D. More formally, the ith allocation of P is deterministic if $P_{i,P}\{A|I_{i-1}(D)\}$ is 0 or 1. At the other extreme, the ith allocation is unpredictable if and only if the prior and posterior distributions coincide, that is, if $P_{i,P}\{A|I_{i-1}(D)\} = P_{i,P}\{A\}$. Otherwise, the allocation is predictable. Note that this does not imply that the allocation is deterministic; it means only that one can predict it with better chances than would be possible based on only knowledge of the allocation procedure P. If P is symmetric, then $P_{i,P}\{A|0\} = 0.5$.

Conditions T and B have implications for the prediction of allocations. In particular, Condition B implies that $P_{i,P}\{A|b\} = 0$ and $P_{i,P}\{A| - b\} = 1$. Condition B applied with $b = 1$ forces each even-numbered allocation to be deterministic; with $b = 2$ (block size 4), every fourth allocation (and sometimes the one just before it) is deterministic. Also, Condition T implies that $P_{i,P}\{A|2N - i\} = 0$ and $P_{i,P}\{A|i - 2N\} = 1$, so any allocation for which $|I_k(D)| = 2N - k$, including the $2N$th allocation, is deterministic. With unrestricted randomization, no allocations are deterministic, or even predictable. **Condition F** specifies 1:1 balance, or $I_{2kb}(D) = 0$, for $k = 1, 2, \ldots, N/b$. This condition is essentially the mechanism by which the randomized block design achieves its terminal balance, Condition T, and its ongoing balance, Condition B. Yet while Condition F may imply both Condition T and Condition B, it is not implied by their combination, and so it may be a stronger condition than is needed. For the randomized block procedure, the allocation number i can be represented as $i = 2k(i)b + h(i)$ for some block number $k(i) \in \{0, 1, 2, \ldots, (N/b) - 1\}$ and block position $h(i) \in \{1, 2, \ldots, 2b\}$. With this representation, we find that one consequence of Condition F is that certain allocations are predictable, and others are deterministic. Specifically, any allocation made when there is not perfect balance is predictable (because the treatment so far less well represented is more likely to be allocated next), and $P_{i,P}\{A|2b - h(i)\} = 0$ and $P_{i,P}\{A|h(i) - 2b\} = 1$ (Follmann and Proschan, 1994).

As in Section 5.3.1, smaller block sizes result in greater balance over the course of the trial, but at the expense of more prediction. In Section 5.3.1 we mentioned varying the block sizes as one possible solution to the problem of optimal selection of block sizes. Having defined notation, we are now in a position to pursue this option.

5.3.3 Varying the block sizes

The variable block procedure still uses blocks, and still forces returns to perfect balance at the end of each block, but in theory these block sizes would not be known, even if the randomization procedure itself is known, because it would specify only the mechanism for selecting the block sizes, but not the block sizes themselves. We now see a misunderstanding similar to the one surrounding the precise meaning of masking and allocation concealment. Specifically, there seems to be a common perception that if the block sizes are varied, and not overtly revealed, then they remain unknown, and therefore no prediction is possible. However, as Rosenberger and Lachin (2002, Section 6.5.2) point out, the variable block procedure may allow *more* prediction than the randomized block procedure with fixed block size. Berger *et al.* (2003a) confirmed this surprising result with a simulation study, and explained the phenomenon with appeal to a set of conditions, or restrictions, on a randomization procedure.

Condition V specifies that $I_k(D) = 0$, forced returns to perfect balance, at least a fixed number of times during the trial. Each sequence in Γ(VBP), the set of admissible allocation sequences for the variable block procedure, satisfies Condition V. Now the randomized block procedure forces such returns to perfect balance N/b times, once for each block. If the variable block procedure forced this same number of returns to perfect balance, then it would allow some of the larger blocks to have more than $2b$ patients. This would entail a violation of Condition B, which specifies that no block size may exceed $2b$. But if Conditions B and V are imposed together, then there could be no such larger blocks to offset the smaller ones. To satisfy both conditions, the variable block procedure would have to force more returns to perfect balance than the randomized block procedure would. As mentioned earlier, it may seem that these forced returns to perfect balance are not a concern for the variable block procedure, because one would not know the size of the present block anyway. But consider that if $|I_k(D)| = b$, then Conditions V and B together would reveal both the size of the current block and the block position of k.

Suppose, for example, that there are two treatment groups and that $b = 3$, so that the largest block size would be 6. But if we vary the block sizes, then not every block would be of size 6. Any given block size would be 2, 4, or 6. We can specify the probability of each, or we can specify that each block size occurs a given number of times and then randomly permute the order in which these block sizes appear, but these decisions are not relevant to the present discussion. If, at some point during the trial, it is observed that the imbalance is $|I_k(D)| = 3$, then the size of the present block could not be 2 or 4. By the process of elimination, this present block would have to be of size 6. In fact, we know more than this; we also know that we have just completed the third allocation in this block, that there are three allocations left in this block, and that the next three allocations must restore the balance, so they are deterministic.

This mechanism by which future allocations can be predicted even with varying block sizes applied only when the largest block size is encountered and it is the extreme type that maximally separates the treatment assignments, so that either the first half are to A and the last half are to B or vice versa. One would not expect there to be too many of these block types, and in fact one could even eliminate them from consideration. This might be a good step to take, but it would not eliminate the prediction of future allocations, because there is another mechanism by which future allocations can be predicted even when the block sizes are varied.

Even for predictable but not deterministic allocations, when the investigator has a good idea of what is coming next but is not sure of it, he or she can employ the convergent strategy, guessing that the next allocation will be to the less well-represented group (Rosenberger and Lachin, 2002, Section 6.5.2). Inflation of the Type I error rate will depend on how often this guessing strategy is correct. Of course, the convergent strategy can be used whether blocking is used at all, and if it is, whether the block sizes are fixed or varied. But among the blocking procedures with a given MTI, using a fixed block size uses the largest blocks, the fewest blocks, and the fewest restrictions. This means that fixed block sizes minimize both the forced returns to perfect balance and the likelihood of the investigator guessing correctly. Varying the block size forces the convergent strategy to be correct more often than it otherwise would have been, and this is true regardless of whether or not the investigator who is guessing knows that the block sizes have been varied. This is how the variable block procedure can lead to substantial prediction of future allocations, selection bias, and inflation of the Type I error rate.

5.3.4   The maximal procedure

Allocation procedures have been identified (Blackwell and Hodges, 1957; Stigler, 1969) that minimize selection bias against certain guessing strategies, but these procedures generally do not satisfy Condition B or control chronological bias. Conversely, the randomized blocks procedure, which does control chronological bias, allows for substantial prediction, even if the block sizes are varied. This would be troubling if the randomized block procedure were the only option for controlling chronological bias, but fortunately it is not. One promising alternative is the maximal procedure, which takes as input the extent of chronological bias allowed by the randomized block procedure, as measured by its MTI. A smaller MTI forces better balance, but requires more restrictions, and this leads to more prediction.

If the randomized block procedure is proposed, with a given block size, then the MTI is half this common block size. If, however, the block sizes vary, then the MTI is half the largest block size. Either way, there are other procedures that match this MTI, but may be more or less predictable than the randomized block procedure or the varied block procedure. Subject to the MTI constraint, the goal is to find the least restrictive allocation procedure, so as to make the procedure more resistant to selection bias than the randomized block procedure or the varied block procedure is. That is, we wish to constrain the optimization, so that predictability is minimized subject to Conditions T (terminal balance) and B (adherence to the MTI).

Reducing prediction involves increasing the cardinality of Γ, the set of admissible allocation sequences (Berger *et al.*, 2003a). Given Conditions T and B, forced returns to perfect balance (Conditions V or F) do not help control chronological bias; they do, however, increase prediction. The cost (in terms of increased prediction) of the additional restrictions does not buy us anything in terms of reduced chronological bias. The maximal procedure is constructed by placing a uniform distribution on the maximal set of allocation sequences that satisfy Conditions T and B, denoted Γ(MP). That is, the set of admissible allocation sequences consists of those sequences that satisfy Conditions T and B, without regard to Conditions V or F. Dropping these latter restrictions is getting something for nothing. Figure 5.1 of (Berger *et al.*, 2003a) shows the different sequences that are possible when the forced returns to perfect balance associated with randomized blocks are dropped. Table 5.2 of (Berger *et al.*, 2003a) shows the derivation of the number of admissible sequences for the maximal

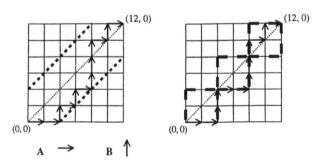

Figure 5.1 Maximal procedure AABABABBBABA (left) with $b = 2$, and an allocation by three randomized blocks of size 4 AABBAABBABAB (right). The diagonal corresponds to perfect balance. Dashed lines are the boundaries for the treatment imbalance allowed by each allocation procedure. Minimizing predictability while retaining balance through the use of less restrictive randomization procedures, Berger, V. W. Ivanova, A. and Knoll, M. D. 2003, copyright John Wiley & Sons, Ltd. Reproduced with permission.

Table 5.2   (a) Number of sequences satisfying MTI = 2 by patients randomized (k) and imbalance. Minimizing predictability while retaining balance through the use of less restrictive randomization procedures, Berger, V. W. Ivanova, A. and Knoll, M. D. 2003, copyright John Wiley & Sons, Ltd. Reproduced with permission

$k =$	0	1	2	3	4	5	6	7	8	9	10	11	12
2	0		1(AA)	0	**3**	0	**9**	0	27	0	81	0	243
1		**1(A)**	0	**3**	0	**9**	0	**27**	0	81	0	243	0
0	0	0	**2**	0	6	0	18	0	**54**	0	162	0	**486**
−1		1(B)	0	3	0	9	0	27	0	**81**	0	**243**	0
−2	0		1(BB)	0	3	0	9	0	27	0	**81**	0	243

(b) Number of sequences satisfying MTI = 3 by patients randomized (k) and imbalance

$k =$	0	1	2	3	4	5	6	7	8	9	10	11	12
3	0	0		1	0	4	0	14	0	48	0	164	0
2	0		1(AA)	0	**4**	0	**14**	0	48	0	164	0	560
1		**1(A)**	0	**3**	0	**10**	0	**34**	0	116	0	396	0
0	0	0	**2**	0	6	0	20	0	**68**	0	232	0	**792**
−1		1(B)	0	3	0	10	0	34	0	**116**	0	**396**	0
−2	0		1(BB)	0	4	0	14	0	48	0	**164**	0	560
−3	0	0		1	0	4	0	14	0	48	0	164	0

For illustration only, the path of the sequence ABAABABBBBAA (chosen arbitrarily) is bolded.

procedure. Table 5.3 of (Berger *et al.*, 2003a) shows the number of admissible sequences for the maximal procedure and the randomized block procedure.

The maximal procedure bears some resemblance to the big stick rule (Soares and Wu, 1983) and the biased coin design with imbalance intolerance (Chen, 1999), except that the transition probabilities are different. See Rosenberger and Lachin (2002, Section 3.6).

Predictability, and therefore selection bias, can be reduced with a corresponding reduction in the number of restrictions on the randomization. When using the randomized block procedure, this would translate into increasing the block sizes. Unfortunately, this may lead to a greater tolerance for imbalance, and therefore to chronological bias. The reduction in predictability achieved by the maximal

Table 5.3 Number of sequences (for the RBP/MP) with varying N and MTI. Minimizing predictability while retaining balance through the use of less restrictive randomization procedures, Berger, V. W. Ivanova, A. and Knoll, M. D. 2003, copyright John Wiley & Sons, Ltd. Reproduced with permission

N (the number of patients allocated to each treatment group, or half the sample size)

MTI	1	2	3	4	5	6	7	8	9	10
1	2/2	4/4	8/8	16/16	32/32	64/64	128/128	256/256	512/512	1024/1024
2	2/2	6/6	12/18	36/54	72/162	216/486	432/1458	1296/4374	2592/13122	7776/39366
3	2/2	6/6	20/20	40/68	120/232	400/792	800/2704	2400/9232	8000/31520	16000/107616
4	2/2	6/6	20/20	70/70	140/250	420/9001	400/3250	4900/11750	9800/42500	29400/153750
5	2/2	6/6	20/20	70/70	252/252	504/922	1512/3404	5040/12630	17640/46988	36504/175066

procedure relative to the randomized block procedure does not require a corresponding increase in imbalance (although the reduction in predictability achieved by the maximal procedure with a larger MTI relative to the maximal procedure with a smaller MTI does require a corresponding increase in imbalance). In fact, the maximal procedure offers the maximal reference set subject to the MTI constraint, Condition B.

Let $n(\text{MP}_{b,N})$ and $n(\text{RBP}_{b,N}) = [(2b)!/(b!)^2]^{N/b}$ be the number of admissible sequences in $\Gamma(\text{MP})$ and $\Gamma(\text{RBP})$, respectively. Because all $D \in \Gamma(\text{RBP})$ satisfy T and B, $\Gamma(\text{RBP}_{b,N}) \subset \Gamma(\text{MP}_{b,N})$ and $n(\text{RBP}_{b,N}) \leq n(\text{MP}_{b,N})$ for all pairs (b,N). If $b = 1$, then each even-numbered allocation needs to be to the treatment group other than the one to which the preceding allocation was directed. In this case, the maximal procedure reduces to the randomized block procedure with N blocks of size 2 each, $n(\text{MP}_{1,N}) = 2^N$. Likewise, if $b \geq N$, then we have a situation similar to when the shot clock is turned off in basketball because there is less time remaining in the quarter than there is on the shot clock. Here, the restriction imposed by Condition B is redundant given Condition T, because if $b \geq N$, then the imbalance could never exceed b anyway. In this case, Condition T is the only restriction, and so the maximal procedure reduces to the randomized block procedure, $n(\text{MP}_{N,N}) = (2N)!/(N!)^2$, with one large block of size $2N$.

For any N, only the two maximally separated sequences, for which $|I_N(D)| = N$ (specifically, treatment A for the first N patients and then treatment B for the last N, denoted $A^N B^N$, and treatment A for the first N patients and then treatment B for the last N, denoted $B^N A^N$) are inadmissible for the maximal procedure when $b = N - 1$, so $n(\text{MP}_{N-1,N}) = [(2N)!/(N!)^2] - 2$. Table 1 of Berger *et al.* (2003a) illustrates the use of Pascal triangles for a systematic derivation of $n(\text{MP}_{b,N})$ for $N \leq 6$ and $b = 2, 3$, as well as boundaries and the paths of any of the admissible sequences. Berger *et al.* (2003a) showed that the maximal procedure has fewer deterministic allocations than the randomized block procedure, especially for larger sample sizes. Berger *et al.* (2003a) also showed, by way of a simulation, that the maximal procedure allows for less inflation of the Type I error rate than the randomized block procedure, with fixed or varying block sizes. This comparison did not take into account two additional ways in which the maximal procedure can lead to less prediction of future allocations than the corresponding randomized block procedure. Table 5.4

Table 5.4   Actual type I error rates of the RBP, variable block (VB) and maximal procedure (MP), $2N = 24$ hypothetical patients, with varying mean response (Δ), 10 000 runs. Minimizing predictability while retaining balance through the use of less restrictive randomization procedures, Berger, V. W. Ivanova, A. and Knoll, M. D. 2003, copyright John Wiley & Sons, Ltd. Reproduced with permission

	$G = 0.5$			$G = 0.99$		
	RBP	VB	MP	RBP	VB	MP
$b = 2$						
$\Delta = 0.0$	0.05	0.05	0.05	0.05	0.05	0.05
$\Delta = 0.5$	0.27	0.29	0.22	0.21	0.11	0.13
$\Delta = 1.0$	0.65	0.70	0.53	0.50	0.19	0.25
$\Delta = 1.5$	0.91	0.94	0.80	0.77	0.31	0.41
$\Delta = 2.0$	0.99	0.99	0.93	0.93	0.43	0.57
$b = 3$						
$\Delta = 0.0$	0.05	0.05	0.05	0.05	0.05	0.05
$\Delta = 0.5$	0.23	0.26	0.17	0.15	0.08	0.09
$\Delta = 1.0$	0.55	0.63	0.40	0.34	0.13	0.15
$\Delta = 1.5$	0.83	0.89	0.64	0.57	0.19	0.24
$\Delta = 2.0$	0.95	0.98	0.80	0.77	0.26	0.33
$b = 4$						
$\Delta = 0.0$	0.05	0.05	0.05	0.05	0.05	0.05
$\Delta = 0.5$	0.21	0.24	0.16	0.12	0.07	0.08
$\Delta = 1.0$	0.49	0.61	0.35	0.26	0.10	0.13
$\Delta = 1.5$	0.76	0.88	0.56	0.43	0.14	0.19
$\Delta = 2.0$	0.90	0.97	0.72	0.60	0.18	0.26

(Berger *et al.*, 2003a) shows the Type I error rate inflation from the maximal procedure, the randomized blocks procedure with fixed block size, and the randomized blocks procedure with varying block size.

First, while prediction is still possible with the maximal procedure, the deterministic allocations for the maximal procedure tend to occur at allocations other than the ones for which investigators are accustomed to guessing (those that are deterministic for the randomized block procedure). Trying to bias patient selection by predicting upcoming allocations involves both effort and risk, and so if enough early attempts to engage in selection bias are thwarted, then a researcher who was otherwise inclined to try to predict allocations might sense a futility in the endeavor, and stop trying, resulting in the complete elimination of selection bias. Second, with the randomized block procedure, one could lose track of the position of the trial, and then pick it up again later, knowing that the imbalance must be zero at certain points during the trial. For the maximal procedure, however, one would need to know just about all of the prior allocations, and certainly not just one or two, to know the current location $\{i - 1, I_{i-1}(D)\}$ of the random walk, and therefore be able to predict where the path is going. This means that if there is only partial unmasking in a masked trial, then the advantage of the maximal procedure is even larger.

The maximal procedure was recently used in a randomized trial in advanced non-small cell lung cancer (Socinski *et al.*, submitted). The rationale for using the maximal procedure, instead of randomized blocks, was the inability of the randomized blocks procedure to adequately control both chronological bias and selection bias.

5.3.5 Extensions

For many combinations of sample sizes N and maximum tolerated imbalance level b, $n(\text{MP}) > 2n(\text{RBP})$. When this is the case, the set of admissible allocation sequences for the maximal procedure, even stripped away of the admissible allocation sequences for the randomized block procedure, will still be larger than the set of admissible allocation sequences for the randomized block procedure that were stripped away. As such, one could prevent a predictable repeating block sequence from occurring by chance and use as the reference set $\Gamma(\text{MP}) - \Gamma(\text{RBP})$, and still have a large enough permutation sample space to reduce conservatism and increase the power of design-based permutation tests (Berger *et al.*, 2003a). This might be a promising approach in larger trials. The maximal procedure can also accommodate uneven allocation, say 3:4, by adapting Conditions T and B to reflect the ideal ratio. Specifically, Condition T would now specify that the final sample sizes in the groups are in exactly the ratio specified, and Condition B would limit deviations from this ratio.

Prediction would be more difficult if Condition T were dropped and deviations from the ideal (for example, 1:1) ratio were tolerated. Note that Condition T is not redundant given Condition B, but nevertheless in the presence of Condition B we find that Condition T offers limited additional benefit. The maximal procedure could easily handle this modification by using as the set of admissible allocation sequences those that satisfy Condition B, without regard to Condition T. Another use of the maximal procedure is to facilitate the use of larger blocks (and, hence, less prediction) with the randomized block procedure. To see how this would work, note that generally there are no restrictions within any block other than balance at the end, a within-block version of Condition T. Using Condition T as the only restriction has been called the random allocation rule (Schulz and Grimes, 2002b). Of course, if the block size is too large, then Condition B would be violated. But we could use excessively large blocks and within them replace the random allocation rule with the maximal procedure (Condition B, with or without Condition T), to ensure adherence to the desired MTI.

Instead of imposing hard boundaries that cannot be crossed, as the maximal procedure does, one could penalize paths proportionally to the imbalance they allow. Specifically, as proposed by Berger *et al.* (2003a), one would enumerate all allocation sequences D (possibly satisfying Condition T) and for each one record $I(D)$, the largest attained imbalance. Now the probability of selecting any given sequence D is $P\{D\} = z^{\max(I(D)-b,0)}/Z$, where z is chosen from the unit interval $[0,1]$ to indicate a level of tolerance for imbalance, $Z = \sum_{D \in \Gamma} z^{\max(I(D)-b,0)}$, and 0^0 is defined to be 1. Notice that $z = 0$ yields the maximal procedure with MTI $= b$, $z = 1$ yields unrestricted uniform randomization, and for intermediate values of z we encourage but do not force balance throughout the trial by penalizing sequences proportionally to the maximum imbalance they allow. If $z \geq 0$, then there are no deterministic allocations.

As discussed earlier in this chapter, there may be emergencies that either actually require knowledge of the intervention that was received or appear to require knowledge of the intervention that was received. Either way, this could lead to intentional unmasking of the allocation for the patient in question. If the randomized block procedure is used, and the block in which the unmasked patient was randomized is not yet complete, then Berger and Exner (1999) suggested that patients stop being recruited into this block, thereby leaving this block incomplete. This is probably an important step to take in reducing selection bias, but it could compromise adherence to the MTI and lead to chronological bias. This presents less of a problem for the maximal procedure, however.

6

Detecting Selection Bias in Randomized Trials

We have seen that even in properly randomized trials with allocation concealment, baseline imbalances in important covariates may
occur. Moreover, these imbalances may be systematic, and may not
reflect chance alone, as they would persist in future trials conducted
at the same institutions by the same investigators, provided that the
same subversions took place. These imbalances can lead to misleading conclusions, regardless of their origin (Vamvakas, 2000), and so
it may not be immediately obvious that there is any benefit in classi-
fying baseline imbalances as random or systematic. Indeed, Rothman
(1977) stated that 'What matters in a particular trial is whether confounding in this trial is present and, if so, to what extent. It is of no
interest to learn whether the confounding which exists might be compatible with chance as to its etiology.' While we might have agreed with
this position prior to the development of specialized methods to deal
with selection bias in randomized trials, we will now have to disagree
with it. The methods to correct for confounding (baseline imbalances
in important covariates) attributable to selection bias differ from the
methods to correct for confounding attributable to random variation.
Briefly, the confounding caused by an unbalanced measured covariate is corrected by simple adjustment for this covariate. Clearly, there
is no way to adjust for a covariate, even if it is both prognostic and
observable, if it is not recorded.

If other remedial methods exist to correct for unbalanced latent
covariates, then it is worth knowing when there are unbalanced
latent covariates. We cannot offer any methods to detect *randomly*

Selection Bias and Covariate Imbalances in Randomized Clinical Trials   V. W. Berger
© 2005 John Wiley & Sons, Ltd.

unbalanced latent covariates, but there are methods to detect unbalanced latent covariates that are the result of third-order residual selection bias. These methods include an assessment of the cause, random or systematic, of imbalance among the measured covariates. After all, it seems reasonable to suppose that evidence of systematic imbalances in measured covariates is more suggestive of systematically unbalanced latent covariates than random imbalances in measured covariates are.

Because of this, it is important to detect third-order residual selection bias when it is present. In other words, 'Since . . . the trialist not only treats but allocates we must have some way of satisfying ourselves that it is the treatment and not the allocation which brings about the effect. On this interpretation randomization is a necessary but not sufficient guarantee for probabilistic calculations' (Senn, 1991). In developing methods to detect third-order residual selection bias when it is present, we will consider the precise mechanism of third-order residual selection bias, and the role played by the selection covariate, the primary covariate, and the probability of allocation to the experimental treatment P{E}, which for reasons to be explained we also call the reverse propensity score (RPS). Recall that P{E}, or the RPS, is the probability, conditional on all previous allocations and the allocation procedure (restrictions on the randomization), that a given patient will receive a given treatment. We first note that in the absence of third-order residual selection bias, certain patterns in the data would not be expected. We exploit this fact to develop formal tests, based on these unexpected patterns, that can be used to detect third-order residual selection bias.

6.1 BASELINE IMBALANCES IN OBSERVED COVARIATES

One pattern that would not be expected in the absence of third-order residual selection bias is a disproportionate number of substantial baseline imbalances. It has become popular to denounce baseline testing within the context of randomized trials as illogical, based on the supposition that randomization by itself guarantees that any imbalance must necessarily be random. If this were true, then the baseline test would be of a null hypothesis that were known to be true, so any rejection of this null hypothesis would represent a Type I error (Senn, 1994). In fact, Rothman (1977) states that 'For clinical trials with proper randomization, any confounding that exists will be the result of chance. In such an instance, statistical significance tests amount to a futile exercise in verifying the adequacy of the randomization process.' But in fact, as we have seen, the premise is not true. If a covariate is unbalanced beyond what can be explained by chance, then this suggests some other mechanism for the imbalance, even if some theory suggests that this should not be the case, or at least should not be the case very often.

Recognizing the possibility of flawed randomization, Burgess, Gebski, and Keech (2003) offer a somewhat more enlightened position on baseline testing compared to those who would abolish it altogether, as they state that 'Use of statistical tests to compare the balance and/or values of baseline characteristics between the study groups and the presentation of p-values are not uncommon. However, many authors assert that this is inappropriate. If randomization has been performed correctly, chance is the only explanation for any observed difference between groups at the outset of the study, in which case statistical tests become superfluous. Consequently, only if it is suspected that the randomization process has failed or was flawed, can performing significance tests on the baseline data be readily justified.' While confining the use of baseline tests to situations in which flawed randomization is suspected is preferable to never using it, does this approach go far enough? It seems to miss the fact that the basis for the suspicion of flawed randomization can be the very tests of baseline balance that would not be performed, under this approach, without *prior* suspicion.

The parallel with the logic of efficacy testing is obvious. A significant treatment effect is generally claimed if the primary efficacy analysis has a sufficiently low p-value, because if chance is ruled out in a probabilistic sense, then the attribution of the observed results must be to the treatment effect (Senn, 1997, Section 7.2.1). For some reason, parallel logic is not applied very often to baseline testing, in which low p-values could again be taken as evidence of flawed randomization. While the literature condemns the practice of formal testing for baseline balance, fortunately this practice remains fairly common in practice, although it is unlikely that the baseline tests are conducted for the right reason. That is, it is likely that many practitioners who conduct baseline tests do so not out of the enlightenment of recognizing selection bias as a potential explanation, but rather to ascertain if an unlucky realization of the randomization sequence was obtained. It is immaterial whether or not practitioners recognize the distinction between the accession numbers in one group being a random sample of all used accession numbers and the patients themselves in one group forming a random sample of all randomized patients.

The former statement would be known to be true in a properly randomized trial, whereas the latter would not. But even if the misunderstanding described by Senn (1997, Section 7.2.1) is behind the continued use of baseline testing, nevertheless these calculations remain useful, even if they are useful for a reason that is lost on those who are performing these very calculations. When statistically significant baseline imbalances are found, it is common to compare the number found to the number that would be expected by chance alone. This comparison may be useful. Indeed, one low baseline p-value out of a large number of covariates tested would not, by itself, rule out chance as a cause. Still, this comparison has its limitations, because clearly there are patterns of baseline imbalances, even if only a few of them among a large set of covariates, that would tend to raise suspicion. For one thing, if there are 20 covariates and only one baseline p-value is significant at the 0.05 level, then this may not be a cause for concern, unless that single baseline p-value happens to be 0.0001 and not 0.049.

We see that the extent of imbalance in each covariate, and not simply the number of unbalanced covariates at an arbitrary level such as 0.05, is quite relevant. In fact, without third-order residual selection bias, the baseline covariate p-values should appear to be uniformly distributed on the unit interval (Altman, 1985), meaning that the proportion of baseline p-values falling under any given value, 0.05 or otherwise, should be roughly that value. If for some value k, with $0 < k < 1$, substantially more than $100k\%$ of the baseline p-values are less than k, then this suggests selection bias.

That the null distribution of a p-value deviates from uniform on the unit interval only because of discreteness (Berger, 2001) is true regardless of whether that p-value is one-sided or two-sided. Generally, baseline p-values are two-sided, but one-sided baseline p-values may be more appropriate, because it is worth knowing if all or most of the imbalances go in the same direction. That is, if a covariate is predictive of a response variable, then some values of this covariate are more advantageous (in terms of bringing about good responses) than others. If the advantage conferred by the numerous unbalanced covariates all favour the same treatment group, then this is more suggestive of third-order selection bias than a pattern of imbalances with roughly equal numbers favoring each treatment group. See Sections 3.3.2, 3.3.27 and 3.3.29.

In fact, if the same group is consistently favored by the imbalances, then this suggests selection bias regardless of the concordance or discordance between the specific group that is favored and the vested interests the investigators appear to hold or not to hold. For example, if the investigators appear to have an incentive to demonstrate the efficacy of the active treatment, and each covariate imbalance favors the active treatment, then this might raise suspicion. But even if each covariate imbalance favored the control treatment, this should still raise suspicions, for at least two reasons. First, not all vested interests are malicious. Even investigators with financial motives may act against these motives if they perceive that these contrary actions are in the best interests of their patients. There might then be an incentive to provide the sickest patients with the novel therapy, based on the presumed belief that these are the patients who need it most. For example, there was suspicion of selection bias in the Canadian National Breast Cancer Screening Study (Tarone, 1995; Boyd, 1997), even though the study was designed to evaluate mammography and it was the mammography arm that had the larger proportion of advanced breast cancer detected at baseline by physical exam.

There is also a second reason to consider third-order selection bias even if the imbalances work against the active treatment, but this applies to continuous covariates only. Imagine a trial of a drug to treat a chronic condition, in which all patients are essentially of the same severity but this severity varies over time within each patient, as part of the natural history of the disease. In such a case, one might expect that regression to the mean (Morton and Torgerson, 2003) would, even in the absence of any treatment effect, cause those patients who appear to have the most severe disease at baseline to improve the most. That is, they have the most room for improvement. In such a case, if the analysis plan called for adjustment for baseline severity by analyzing not the post-treatment severity but rather the change from baseline, then there would be a distinct advantage to whichever treatment group had the most patients with severe baseline disease. That is, the apparent advantage for the control group, with the healthier patients at baseline, can ultimately turn into an advantage for the active treatment group by virtue of the fact that the adjustment achieved by analyzing the change from baseline overcompensates for the baseline imbalance in the covariate disease severity. This concern does not apply to nominal or binary covariates, because adjustment for these stratification variables consists in comparing the treatment groups within each level of these covariates. With this type of model-independent adjustment, no overadjustment is possible.

In light of the aforementioned concerns, one test for third-order selection bias is based on how uniformly (on the unit interval) the one-sided p-values as an entire group appear to be. As Altman (1985) points out, 'A more powerful approach is to consider the actual p-values observed for all baseline comparisons'. It would be nice if the converse were true, so that having the set of baseline p-values appear to be uniformly distributed on the unit interval suggested a lack of selection bias. We will now explore this converse. It is not common to specify a primary covariate, but imagine if it were (nearly) universally accepted that a certain covariate were the most prognostic of all covariates for the outcomes of interest. For example, Gleason (1977) stated that among prostate cancer patients, 'stronger and more regular correlation was found between mortality rates and this averaging pattern score ... than with any other histological scale'. The pattern score might then be taken to be the primary covariate.

If, in a randomized study of prostate cancer patients, only a few of the many baseline covariates are unbalanced across treatment groups, then this may not be sufficient to ensure valid treatment comparisons if the pattern score (now called the Gleason score) is one of the unbalanced covariates. Moreover, if hundreds of such trials are examined, each with roughly 20 covariates, and in each one the Gleason score is unbalanced, then this would suggest that more than chance is at play, even if it is the only unbalanced covariate. With this in mind, it would seem that an examination into the potential for third-order residual selection bias would not be complete without special attention afforded to particularly predictive covariates, such as the Gleason score. More generally, suppose that the covariates could be ranked by predictive ability for the primary response variable. Certainly, it is also possible to rank the covariates by the degree of imbalance (either absolute magnitude or the p-value). These two rankings, one by predictive ability and one by imbalance (one-sided), should be checked for correlation with each other. If they are correlated, then this would suggest that there is more than chance at play, and perhaps there is third-order selection bias. It is also possible to investigate the cause of a single baseline imbalance, random or selection bias, by using the flow diagram in figure 2.6.1.

6.2 TESTING FOR SELECTION BIAS WITHOUT BASELINE ANALYSES

We see that finding no more than 5% of the covariates to be unbalanced at the 5% level does not ensure a lack of selection bias because those covariates that are unbalanced at the 5% level may have p-values that are much lower than 0.05, and/or may be the most prognostic covariates. There is also a third reason to refrain from excessive excitement when finding that no more than 5% of the covariates are unbalanced at the 5% level, specifically the potential failure to measure some prognostic and unbalanced covariates (Green and Byar, 1984; Moses, 1995). It may seem that there is no way to study potential imbalances in latent covariates, but in some cases there is such a method. Indeed, we offer no methods for identifying unbalanced latent covariates. However, the presence of unbalanced measured covariates may indicate also the presence of unbalanced latent covariates if the imbalances in the measured covariates appear to be the result of third-order selection bias, and not the result of chance alone. Of course, we have not yet discussed how to distinguish random observed baseline imbalances from systematic observed baseline imbalances. We take up this issue now.

6.3 THE SELECTION COVARIATE

The ultimate covariate is the set of potential responses a given patient would have to each treatment under study (Greenland and Robins, 1986; Frangakis and Rubin, 2002). In general, this ultimate covariate is not observed, but it can be estimated or predicted. We define the selection covariate as the estimated ultimate covariate, at the time a decision is made to enroll a screened patient or deny enrollment to this patient. The selection covariate may be measured, or it may be based in part on observable but unmeasured patient characteristics, in which case it is unmeasured. We formulate the mechanism of third-order residual selection bias to have a causal pathway that includes this selection covariate, which is used to form the RPS intervals that are used to make enrollment decisions. For example, the decision may be to enroll the subject if both the selection covariate and the RPS are large or both are small, but not otherwise, or if the RPS falls within the RPS interval of a candidate subject. Third-order residual selection bias represents an effort to create an imbalance in this selection covariate. The effort is successful to the extent that the RPS predicts the actual treatment assignments and to the extent that discretion can be exercised in selecting subjects with specific values of the selection covariate to match the RPS value. This means that strong evidence of third-order residual selection bias serves also as strong evidence of the existence of an unbalanced selection covariate. If this is the case, yet none of the measured covariates are unbalanced, then the selection covariate has not been identified, and must necessarily be latent. By 'latent' we mean observed but not recorded, as opposed to not observed at all.

The concern that latent covariates may be unbalanced even when there are no unbalanced measured covariates is not so far-fetched, considering that latent covariates, such as subjective health perceived by a patient, can predict clinical outcomes and even mortality, even after adjusting for other observed predictors (Fayers and Sprangers, 2002). Hence, proper detection of third-order selection bias requires a method that is not based exclusively on the extent of baseline imbalances in the observed covariates. We now discuss such methods. Some specific methods to detect third-order selection bias utilize the RPS and the screening log.

6.4 THE ROLE OF THE REVERSE PROPENSITY SCORE IN THIRD-ORDER RESIDUAL SELECTION BIAS

Recall that the mechanism for third-order selection bias is to first note that the RPS serves as the predicted treatment assignment, and hence would be expected to differ across treatment groups. That is, similar to a situation described by Heckman *et al.* (1996), both the support and the distribution of the RPS differ across the treatment groups. Knowing this, an investigator intent on inducing confounding would try to correlate the selection covariate with the RPS by selecting patients with large values of the selection covariate when the RPS value is large, and by selecting patients with small values of the selection covariate when the RPS value is small. How large and how small depend on the circumstances. To the extent that the RPS predicts the actual treatment assignment and to the extent that patients can be selected so as to ensure the high correlation of the selection covariate with the RPS, the selection covariate will be unbalanced across treatment groups, again with both different supports and distributions across the treatment groups.

Strong associations between the RPS and any covariate would suggest not only third-order selection bias but also that the associated covariate was, or contributed to, the selection covariate. If the selection bias is 'unobservable' (using the terminology of Berger and Exner, 1999), then the selection covariate is latent, and is not strongly associated with any of the observed covariates (which themselves may be balanced or unbalanced). Note that our method for deciding if a given imbalance in a given covariate is random or selection bias is to evaluate the correlation of that covariate with the RPS. This may constitute progress, but it still does not tell us how to detect unobservable selection bias. Recall that when the selection covariate is latent, its identity is known to the investigator but not to any reviewer of the study, as it is observed but not recorded. In this case, a reviewer would need to think like the investigator, and ask how to best induce confounding with such a latent selection covariate.

From the perspective of the investigator trying to induce confounding, the best patient selection strategy would be to ensure that the selection covariate would be maximally correlated with the RPS. For example, if the selection covariate takes three values, say low, medium, and high, and if randomized blocks are used with a fixed block size of 2, then the RPS will also take on three values, specifically 0.0, 0.5, and 1.0. The patient selection would then be to select low, medium, or high as the RPS value is 0.0, 0.5, or 1.0. Of course, if the selection covariate has more or fewer categories than the RPS, then such a one-to-one correspondence cannot be established. One might wish to take the conservative approach of defending against the worst-case scenario in which a one-to-one correspondence can be established between the RPS levels and the covariate levels. In this case, we do not need to know the selection covariate, because it is functionally equivalent to the RPS itself. This means that the RPS can itself be used as a covariate to separate the effects of selection bias and the intervention.

The RPS is an unusual covariate. Senn (1997, Section 7.1) described three types of covariates, specifically demographic characteristics, outcome measures taken at baseline, and what are essentially time-varying correlates other than the primary outcome. The RPS does not fall into any of these categories, as it is not an inherent characteristic of the patient, and in fact becomes associated with the patient only when randomization of this patient occurs. Once the patient is randomized, the RPS value (for that patient) will not change, so the RPS is not time-varying, nor is it an outcome measure. Yet to the extent that there is third-order residual selection bias, the RPS is informative about the patient characteristics, as it conveys information about the circumstances (likelihood of allocation to either treatment group) under which it was deemed appropriate to randomize this patient. That is, knowing that the patient was selected to be randomized at this particular RPS value tells us, in the presence of third-order residual selection bias, something about how well the investigator thought the patient would respond. In the presence of unobservable third-order residual selection bias, the RPS may well be an independent predictor of outcomes, even above and beyond the prediction possible with other covariates (with which it is not associated). The question remains, though, how can we best use the RPS to determine if third-order selection bias is present?

6.5 USING THE REVERSE PROPENSITY SCORE TO TEST FOR SELECTION BIAS: THE BERGER–EXNER TEST AND GRAPH

Associations of the RPS and response variables are to be expected if the treatments differ in their effectiveness. This is because, with or without third-order residual selection bias, the RPS is associated with the actual treatment assignment by virtue of the restrictions on the randomization (there is no such association if unrestricted randomization is used), and the actual treatment effects may be expected to be associated with the response variables. However, Berger and Exner (1999) recognized that *within a treatment group*, the RPS should not be associated with any response variable. Consider that if the RPS is associated with a response variable within a treatment group, say the active treatment group, then this suggests that the mechanism of action of the active treatment has, on its causal pathway, the likelihood with which it was to be assigned. That is, consider two identical clones, each to be treated for the same disease. One is given the active treatment. The other has his treatment determined by the toss of a coin.

Now suppose that by chance alone, the coin turned up heads, and so this clone also was treated with the same active treatment. Having removed all sources of variation other than the process by which the treatment decision is reached (that is, the RPS), any difference in the outcomes of the two clones would have to be attributable to the RPS itself. There are at least two possible mechanisms by which the RPS could exert such an influence. One is third-order selection bias, although this is not possible in the hypothetical case now considered, because identical patients were selected for each 'group'. The other mechanism is observation bias caused by the knowledge that one clone was to receive the active treatment, and so might be scored differently (at least on subjective assessments) than the other clone, whose treatment identity was not known.

Contrary to conventional wisdom, a lack of masking *can* have an effect even on objective endpoints. For example, a lack of masking can lead to increased attention and/or ancillary care for one group, and this in turn can lead to truly healthier patients in that favored group, with the health disparity reflected even in objective endpoints. Nevertheless, in the case of an actual trial using a heterogeneous study population (not the clones that we considered in the hypothetical example), we still consider third-order residual selection bias to be the more plausible attribution if an objective endpoint is associated with the RPS within a treatment group (and within any stratum that was used in the randomization). This is true especially in an unmasked or a single-masked study, because in these cases any observation bias would be expected to be based not on the RPS, but rather on the actual treatment assignment, that would be known. In this case, observation bias would be completely confounded with the differential effects of the treatments, so neither would be identifiable. Yet neither the observation bias nor the effects of the treatments would affect outcomes differentially across RPS values within a treatment group. Even in a double-masked study, third-order residual selection bias requires unmasking, so again observation bias should be confounded more with the treatment effects than with the RPS effect.

The study of the association of the RPS and objective response variables within each treatment group (and within any stratum that was used in the randomization) is the basis for the Berger–Exner test (Berger and Exner, 1999). Logistic regression, the analysis of covariance, the Cochran–Mantel–Haenszel test, Cox regression, or another test may be used as appropriate, depending on the nature of the objective response variable and how many RPS levels there are. In fact, the RPS can even be treated as an ordinal, rather than a continuous, covariate. Berger *et al.* (2004) provided methodology for this type of adjustment. The goal here is to determine if there is third-order residual selection bias (later we will discuss another objective that is based on essentially the same model), so in any of these models the RPS would need to be fitted after the actual treatment variable. The combination of positive findings from the Berger–Exner test and the lack of imbalances among the recorded covariates would provide strong evidence that there is an unbalanced latent covariate, or unobservable third-order selection bias.

The Berger–Exner graph is a nice visual display that can reveal selection bias if it is there, or rule it out if it is not. Like the Berger–Exner test, the Berger–Exner graph requires data at the patient level. For either the Berger–Exner test or the associated Berger–Exner graph, it is important not to mis-specify the block size, because doing so can lead to spurious findings. To demonstrate, consider a randomized trial entitled 'A Double-Blind Clinical Trial in Carcinoma of the Lung of Immunotherapy as an Adjuvant to Surgery Stage I (Non-Small Cell)' (Mountain and Gail, 1981). There were several centers, but we consider only the data from the Seattle Center. Within each center, including Seattle, there were eight strata, defined by age less than 60 years or at least 60 years; squamous or non-squamous; and pneumonectomy or subtotal resection (so $2 \times 2 \times 2 = 8$). BCG (Tice Strain) and Isoniazide (INH) were compared to placebo. We consider only the non-squamous patients with subtotal resection, so this includes two strata (stratum 4 based on age less than 60 years, with 18 patients, and

Table 6.1   Stratum 4, Lung Carcinoma Study (Mountain and Gail, 1981)

PATIENT	TREAT	CENSORED	SURVIVAL	REMAIN	REMAIN_2	R_P_S_4
01	2	1	1863	4	2	0.50000
02	1	0	4186	3	1	0.33333
03	2	0	4157	2	1	0.50000
04	1	0	4099	1	0	0.00000
05	1	0	3776	4	2	0.50000
06	2	0	3969	3	2	0.66667
07	1	1	3823	2	1	0.50000
08	2	1	2541	1	1	1.00000
09	2	1	743	4	2	0.50000
10	1	1	530	3	1	0.33333
11	1	1	379	2	1	0.50000
12	2	1	698	1	1	1.00000
13	1	1	207	4	2	0.50000
14	2	1	1620	3	2	0.66667
15	2	0	3316	2	1	0.50000
16	1	0	3190	1	0	0.00000
17	1	0	3100	4	2	0.50000
18	2	1	1968	3	2	0.66667

stratum 8, based on age at least 60 years, with 29 patients). The data for stratum 4 are as follows (note that BCG is coded as Treatment 1, and placebo is coded as Treatment 2):

The documentation for this trial available to this author did not include a block size, but it is quite common for two-arm trials to use blocks of size four (see Section 5.3.1), and so this was used in the calculation of the RPS in Table 6.1. The Berger–Exner graph displayed in figure 6.1

We notice several important features of this graph. First, whichever treatment was more likely to be assigned actually was assigned. Second, when considering intermediate RPS values ($\frac{1}{3}$, $1/2$, and $\frac{2}{3}$), the survival times do not appear to be influenced by either the RPS value or by the treatment group to which the patient was assigned. However, when Treatment 1 was certain to be assigned (RPS = 0), the survival times are all greater than 3000, and when Treatment 2 was certain to be assigned (RPS = 1), the survival times are all less than 3000. This causes the survival times among Treatment Group 1 to be longer when RPS = 0 than when RPS = $\frac{1}{3}$ or $1/2$. Likewise, the survival times

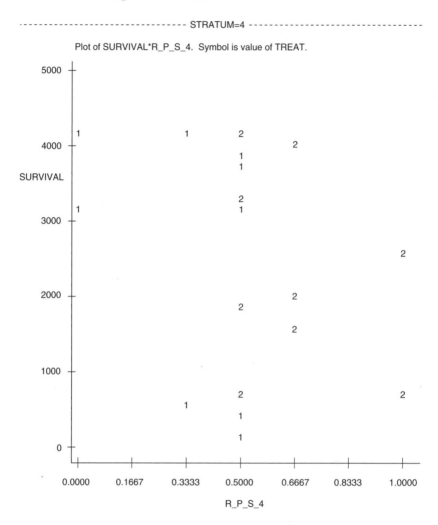

Figure 6.1 Berger–Exner graph for stratum 4 and block size 4, lung carcinoma study.

among Treatment Group 2 patients tend to be shorter when RPS $= 1$ than when RPS $= \frac{2}{3}$ or $^1/_2$. This pattern is suggestive of selection bias, although the sample size is sufficiently small to preclude a definitive conclusion. The same trend emerges in stratum 8, in Table 6.2.

The Berger–Exner graph for Stratum 8 is as follows.

Table 6.2   Stratum 8, Lung Carcinoma Study (Mountain and Gail, 1981)

PATIENT	TREAT	CENSORED	SURVIVAL	REMAIN	REMAIN_2	R_P_S_4
01	2	1	641	4	2	0.50000
02	1	1	616	3	1	0.33333
03	1	1	1577	2	1	0.50000
04	2	1	504	1	1	1.00000
05	2	1	1600	4	2	0.50000
06	1	0	3013	3	1	0.33333
07	1	1	2089	2	1	0.50000
08	2	1	2492	1	1	1.00000
09	1	0	2727	4	2	0.50000
10	2	0	3240	3	2	0.66667
11	1	1	194	2	1	0.50000
12	2	1	457	1	1	1.00000
13	1	0	3780	4	2	0.50000
14	2	0	3900	3	2	0.66667
15	2	1	1583	2	1	0.50000
16	1	1	3041	1	0	0.00000
17	1	1	262	4	2	0.50000
18	2	1	480	3	2	0.66667
19	2	0	3307	2	1	0.50000
20	1	1	3332	1	0	0.00000
21	2	0	3452	4	2	0.50000
22	1	0	3312	3	1	0.33333
23	1	0	3209	2	1	0.50000
24	2	1	39	1	1	1.00000
25	1	0	3289	4	2	0.50000
26	2	1	443	3	2	0.66667
27	1	1	1675	2	1	0.50000
28	2	1	31	1	1	1.00000
29	2	1	1041	4	2	0.50000

The same pattern emerges, with long survival times for RPS = 0 values, short survival times for RPS = 1 values, and intermediate survival times for intermediate RPS values. A stronger pattern emerges when the two strata are combined.

The trend in this graph, with the association between RPS values and survival times within treatment groups, is suggestive of selection bias. Indeed, this might have made a good example for Chapter 3, except that these calculations were all based on a block size of four.

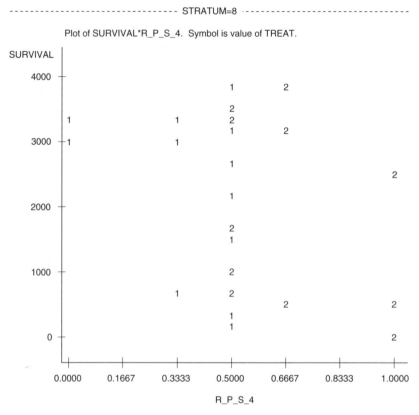

Figure 6.2 Berger–Exner graph for stratum 8 and block size 4, lung carcinoma study.

Again, this author was unable to ascertain what the block size actually was, and a block size of four appeared to be reasonable, except that another pattern is clear in the data, namely that the data are also consistent with a block size of two. This does not, of course, mean that the block size could not have been four, but it is possible to compute the probability of this event if one is willing to assume that the block size could have been only two or four, and the two are equally likely. If the block size is indeed two, then the probability of consistency with a block size of two is 1. If, on the other hand, the block size is four, then the probability of consistency with a block size of two is much less.

Plot of SURVIVAL*R_P_S_4. Symbol is value of TREAT.

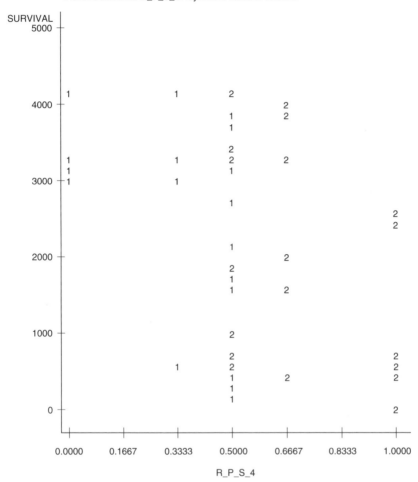

NOTE: 10 obs hidden.

Figure 6.3 Berger–Exner graph for stratum 4 and 8 combined, and block size 4, lung carcinoma study.

There are nine pairs of patients in stratum 4, and 14 pairs of patients in stratum 8 (the 29th patient is part of an incomplete block regardless of the block size, and so does not enter into these calculations). To be consistent with a block size of two, each of these 23 pairs need to be balanced, as in one patient per pair allocated to each treatment group.

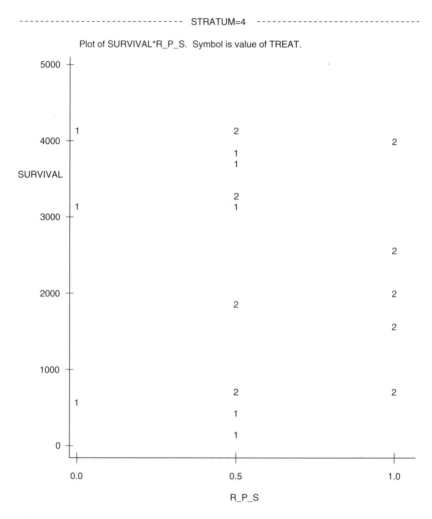

NOTE: 1 obs hidden.

Figure 6.4 Berger–Exner graph for stratum 4 and block size 2, lung carcinoma study.

If the block size is four, then there are six block types (1122, 1212, 1221, 2112, 2121, and 2211). The first and last of these would violate the balance we found, so only four of the six block types are represented, with a probability of $(2/3)^{23}$. By Bayes conditional probability formula, then, the probability that the block size is two given data

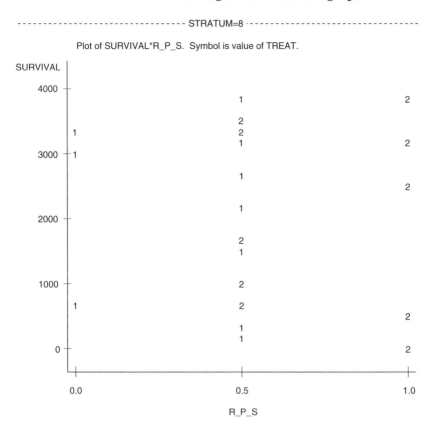

Plot of SURVIVAL*R_P_S. Symbol is value of TREAT.

NOTE: 9 obs hidden.

Figure 6.5 Berger–Exner graph for stratum 8 and block size 2, lung carcinoma study.

consistent with a block size of two is $1/[(2/3)^{23} + 1] = 0.999911$. The probability that the block size is four is negligible, so we evaluate this trial based on the presumption that the block size was two, not four. Now the Berger–Exner graph tells a much different story. First we see that there is no selection bias in stratum 4.

Next we see that there was no selection bias in Stratum 8.

Finally, we see that there was no selection bias when the two strata are combined.

So the mis-specification of the block size is what caused the appearance of selection bias.

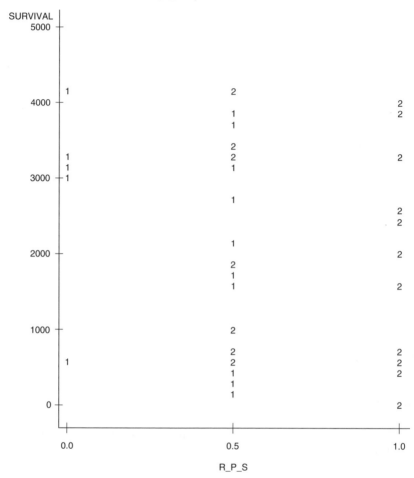

Figure 6.6 Berger–Exner graph for stratum 4 and 8 combined, and block size 2, lung carcinoma study.

6.6 USING THE SCREENING LOG TO TEST FOR SELECTION BIAS

If the screening log is available, as it would be in a comprehensive cohort follow-up study (Olschewski *et al.*, 1992), then it can also be used to provide evidence for or against unobservable selection bias.

For example, the screening log might be used to detect deferred enrollment (as is alluded to in Sections 3.3.9 and 3.3.20), in which screened patients who are not randomized might later be recalled and randomized. Deferred enrollment is hard to justify under any circumstances, but when a given patient is denied enrollment when the RPS has one value, and then randomized later when the RPS takes on a different value, is especially troubling. This 'strategic' deferred enrollment may well signify third-order residual selection bias. A variation on this theme is stratum-distorted enrollment, in which a patient whose covariate values place him or her in one stratum is randomized instead from a different stratum, corresponding to different covariate values. For example, randomization stratified by gender can be accomplished by preparing separate allocation sequences for each gender. At any point during the trial, then, each stratum will have its own RPS value, and these stratum-specific RPS values may well differ across the strata. Stratum-distorted enrollment is strategic if it results in a patient being randomized from an RPS value other than the one that was current (at the time this patient was enrolled) in the proper stratum for this patient. Recall from Chapter 3 that there was strategic stratum-distorted enrollment in the etanercept trial for juvenile rheumatoid arthritis (Section 3.3.13). See also Section 3.3.26.

The screening log can also be used in other ways to try to detect patterns that are suggestive of third-order residual selection bias. For example, with covariate information recorded for not only the randomized patients but also the randomizable (screened but not enrolled) patients, one can cross-classify the RPS and the decision to randomize or not. Then one can consider the mean value of each covariate for each combination of the RPS and the randomization status. To see how such a data display might be useful, recall the CASS Study (discussed in Section 3.3.12), and consider the hypothetical data in Table 6.3, with a single binary covariate within each cell, and blocking with fixed block size 2. The only possible RPS values in this case are 0.0, 0.5, and 1.0. In Table 6.3, 80 patients were screened, of whom 40 were randomized. Of the 80 screened patients, 20 were screened when the RPS value was 0.0 (this would be the second block position after the first patient in the block was randomized to the active treatment), 40 were screened when the RPS value was 0.5 (this would be the first block position), and 20 were screened when the RPS value was 1.0 (this would be the second block position after the first patient in the block was randomized to the control treatment). Overall, 40 of

Table 6.3   Associations among a binary covariate, the RPS, and the randomization status. The reverse propersity score to detect selection bias and correct for baseline imbalances, Berger, V. W., 2005, copyright John Wiley & Sons, Ltd. Reproduced with permission.

	RPS			
	0.0	0.5	1.0	Total
Randomized	0/10	10/20	10/10	20/40
Not randomized	10/10	10/20	0/10	20/40
Total	10/20	20/40	10/20	40/80

the 80 patients were randomized, and 40 of the 80 patients had the preferred covariate value. Notice that neither the randomization status nor the RPS is predictive of (associated with) the covariate when all screened patients are considered. In other words, the randomized group is comparable to the randomizable group, at least with respect to the one covariate considered, as each has 20/40 patients with the preferred covariate value. Furthermore, the groups defined by RPS values are also comparable to each other, as again each is composed of half 'healthier' patients and half 'sicker' patients. This might be taken as evidence of no selection bias; indeed, the similarity (in terms of the covariate distributions) of the randomized and randomizable groups is often mentioned, presumably to make the case that there was no selection bias. And yet the fact that all patients, with good and bad covariate values, can get randomized should not be reassuring.

Table 6.3 shows a clear picture of selection bias, as seen by considering the ability of the combination of the covariate and the RPS (or the interaction term) to predict the randomization status. That is, patients with better covariate values get randomized when RPS = 1, and patients with worse covariate values get randomized when RPS = 0. Specifically, of the 20 screened when the RPS value was 0.0, 10 had the preferred covariate value, and all 10 were denied enrollment. The 10 with the less preferred covariate value were all randomized. Of the 40 patients screened when the RPS value was 0.5, 20 had the preferred covariate value, of whom 10 were randomized; the other 10 were denied enrollment. Likewise, of the 20 with the less preferred covariate value, 10 were randomized and the other 10 were denied enrollment. Of the 20 screened when the RPS value was 1.0, 10 had

Table 6.4   Associations among a binary covariate, the RPS, and the treatment group

	RPS			
	0.0	0.5	1.0	Total
Control	0/10	5/10	0/0	5/20
Active	0/0	5/10	10/10	15/20
Total	0/10	10/20	10/10	20/40

the preferred covariate value, and all 10 were randomized. The 10 with the less preferred covariate value were all denied enrollment. What stands out here is that the pattern of denials when RPS $= 0.5$ is non-selective, as it is not based on the covariate values. Presumably, then, these exclusions are legitimate after all (not all patient exclusions represent a bias). In contrast, the patterns of exclusions when RPS $= 0.0$ or RPS $= 1.0$ make it clear, or at least highly likely, that these exclusions are strategic, and represent selection bias.

The result of the third-order selection bias illustrated in Table 6.3 on the 40 randomized patients is the confounding illustrated in Table 6.4 (and figure 2.6.1). Specifically, 20 healthier patients were enrolled, and 15 of these healthier patients ended up in the active treatment group. Of the 20 sicker patients enrolled, 15 ended up in the control group. So the active group gets three times as many 'good responders' as the control group, 15/20 vs. 5/20. As this example illustrates, given the screening log, one can test for third-order residual selection bias by trying to predict the randomization status with the combination of the covariate and the RPS. This approach, or variations on this theme, was used by Berger and Exner (1999), at the bottom of p. 321 and by Swingler and Zwarenstein (2000), specifically in their Section 2.4 (p. 703).

6.7   THE IVANOVA-BARRIER-BERGER (IBB) DETECTION METHOD

Developed only for use when the randomized block procedure is used, the IBB detection method (Ivanova, Barrier, and Berger, 2005) is based on the presumption that the investigator is aware of both the block sizes and the allocations already made. In this regard, it is similar to the Berger–Exner method discussed in Section 6.5. These two tests differ after this, however. Specifically, the IBB method specifies the existence of strong patients (with response rate p(i) + D), medium patients (with response rate p(i)), and weak patients (with response rate p(i) − D). Here p(i) can be taken as the overall response rate to treatment i (assuming that the population has the same number of strong and weak patients), and $0 \leq D \leq \min\{p(i), 1-p(i)\}$.

The IBB method also specifies a cut-off G such that strong patients are selected for randomization if $P\{E\} > G$, weak patients are chosen if $P\{E\} < 1-G$, and medium patients are selected otherwise (see Section 4.3). The IBB test is a likelihood ratio test of the null hypothesis that D = 0, against the two-sided alternative hypothesis that D is not zero. Clearly, if D is zero, then there is no opportunity for selection bias, because even if an investigator tries to distinguish patients as strong or weak, this will be a fruitless effort. All patients, after all, would have the same response rate. Table 6.5 of Ivanova, Barrier, and Berger (2005) provides the power and Type I error rates of the IBB test of selection

Table 6.5   Power and type I error rates for test of the presence of observable selection bias (H_0: $\delta = 0$) using logistic regression model and linear model

			Cut-off = 0.50		Cut-off = 0.99	
p_1	p_2	δ	Linear model	Logistic model	Linear model	Logistic model
0.50	**0.50**	**0.00**	0.05	0.05	0.06	0.06
0.50	**0.50**	**0.20**	0.95	0.59	0.05	0.59
0.50	**0.70**	**0.00**	0.05	0.05	0.65	0.05
0.50	**0.70**	**0.10**	0.42	0.18	0.68	0.20
0.50	**0.70**	**0.20**	0.96	0.59	0.70	0.60
0.60	**0.40**	**0.00**	0.05	0.05	0.63	0.05
0.60	**0.40**	**0.30**	>0.99	0.98	0.67	0.96
0.70	**0.50**	**0.00**	0.05	0.05	0.65	0.05
0.70	**0.50**	**0.10**	0.44	0.20	0.66	0.20
0.70	**0.50**	**0.20**	0.98	0.70	0.68	0.66
0.70	**0.70**	**0.00**	0.05	0.05	0.06	0.05
0.70	**0.70**	**0.20**	0.9	0.70	0.04	0.67

The data are generated according to model (2). The sample size of 192 is used in each trial.

Table 6.6 Power and type I error rates for test of the presence of observable selection bias (H_0: $\delta = 0$) using linear model (2) and logistic regression model (4)

			Cut-off = 0.50		Cut-off = 0.99	
p_1	p_2	δ	Linear model	Logistic model	Linear model	Logistic model
0.50	**0.50**	**0.20**	0.41	0.37	0.17	0.18
0.50	**0.70**	**0.20**	0.43	0.38	0.17	0.19
0.60	**0.40**	**0.30**	0.78	0.70	0.39	0.40
0.70	**0.50**	**0.20**	0.43	0.38	0.19	0.20

The data are generated according to model of enrolment with limited availability. The sample size of 192 is used in each trial.

bias. Table 6.6 of Ivanova, Barrier, and Berger (2005) provides the power and Type I error rates of the IBB test of selection bias in the more realistic situation in which there is a limited number of strong and weak patients, so that they can be found only half the time.

6.8 INTERPRETING NEGATIVE TESTS OF SELECTION BIAS

It has been suggested (Hollenbeak *et al.*, 2002) that a test for selection bias that fails to reach statistical significance provides evidence that selection bias is not a problem. While we agree with the importance of determining when there is significant evidence of a lack of selection bias, we do not agree that a formal hypothesis test can prove the null hypothesis. That is, a p-value can tell us only if there is or there is not evidence of selection bias. However, the p-value should not be the only assessment of selection bias. Graphing the response variable against the RPS for the different treatment groups (Section 6.5) reveals the evidence of both selection bias and no selection bias. If increasing the level of RPS has a differential effect on the response across treatment groups, then this interaction between treatment and the RPS indicates that when the RPS is large, patients are selected for their potential outcome to the experimental treatment group only. We can further graph the number of patients assigned at each level of RPS and distinguish between different levels of a covariate, such as gender, and within each one see how many are randomized to the intervention group and how many are randomized to the control group. This analysis would tell us whether observing the RPS and having evidence that the next allocation is to the intervention group would influence the choice of the gender of the patient to be randomized. Hence, based on RPS we could detect selection bias towards a certain gender assigned to a specific treatment. If, for example, at each level of RPS, the percentage of males assigned to a specific treatment is the same as the percentage of females receiving the same treatment, then we have evidence of a lack of selection bias.

6.9   WHEN SHOULD ONE TEST FOR SELECTION BIAS?

Is it important to test for selection bias when on the surface there is no evidence to suggest its presence (for example, no substantially unbalanced measured covariates)? As discussed in Section 6.1, Burgess, Gebski, and Keech (2003) suggest that the answer might be no. And yet a negative response to this question might constitute a 'catch-22', as one could argue just as fervently that when there is no evidence that a new drug has efficacy, there is no reason to go through the rigmarole of formally testing the drug for safety and efficacy. And yet, without formally testing the new drug for safety and efficacy, there is no way that there would ever be evidence that the new drug has efficacy. Absence of evidence should never be confused with evidence of absence (Senn, 1997, Section 15.2.1; Barraclough, 2003). There is clearly a benefit in knowing if a given trial had selection bias, because there are methods to correct for this, as we will discuss in Chapter 7. This being the case, to argue that selection bias should not be tested for would require a reason not to do so, and this reason would need to be at least commensurate with the reason to test for selection bias. Perhaps a better question, then, would be 'Under what circumstances would it be reasonable *not* to test for selection bias?'. Let us consider the reasons not to test for selection bias. Note that the debate to test or not to test for selection bias can be cast as a trade-off between progress (the unfettered acceptance of the results of a trial) and caution (the deferment of judgment until all the facts are in).

When viewed in the light of progress against caution, it is not surprising that the efforts to detect selection bias in practice have been so lacking, at least up to this point in time, for 'Precaution is equated with economic and social stagnancy, and is viewed as an unnecessary interference with the scientific advances essential to progress. Progress, as defined by the industrial community, trumps precaution' (Rosner and Markowitz, 2002). Indeed, there is precedent for the existence of a readily available solution to a pressing problem to be insufficient for its use in practice:

> To the modern observer, insisting that doctors wash their hands between patients seems not just obvious, but also benign. But without the 'evidence' that was to come much later, this request seemed to many baseless, and the opposition came from men of science who wanted hard proof of cause and effect. (Kurland, 2002)

Readily available solutions to the problem of silicosis were also resisted (Rosner and Markowitz, 2002), as are readily available solutions (exact analyses that require no distributional assumptions) to the problem of the lack of random sampling from populations with any known distribution in randomized trials (Berger, 2000; Berger *et al.*, 2002). Recall that testing for selection bias might involve analyses of baseline imbalances, with one-sided p-values and rankings of covariates based on how prognostic they were or, preferably, how prognostic they were expected to be.

Recall also that if there is third-order selection bias, then there needs to be a selection covariate, and the trial-specific measure of prognostic ability of each covariate would not be available at the time that the selection covariate would need to be chosen. This is why the expected predicted ability might be preferable, for this purpose, to the observed predicted ability. Of course, it is possible also to modify the choice of selection covariate during the course of the trial, to make use of accruing data regarding prognostic ability of various covariates. Besides analyses of baseline variables, testing for selection bias might also involve analyses of the RPS, including the Berger–Exner test (associating the RPS with response variables within treatment groups and strata) and associating the RPS with covariates within strata. Finally, testing for selection bias might involve scrutiny of the screening log. The baseline variables are available routinely, so very little additional cost or effort would be required to use these as a test for selection bias.

In fact, many trials present baseline p-values, so the first step would simply be to recast these tests as valid tests (although crude ones) of selection bias, rather than as the inappropriate tests for the success of the randomization that they are generally understood to be.

The RPS is not generally presented, or even computed, in actual trials, but certainly the information with which to compute the RPS is readily available. Again, there is very little additional cost or effort required to analyze the RPS for a more refined test of third-order selection bias. The screening log is less ubiquitous, so we might consider separately the cases in which it is and is not available. If the screening log is available, then the analysis of randomization status, RPS value, and covariate values can be automated, as can the search for stratum-distorted enrollment. As for the examination of the screening log for deferred enrollment, it is certainly possible to automate the comparison of the values of all covariates for duplicate records, but examination of personal identifiers, and possibly a manual search for erasure marks or white-out, would be required for a definitive assessment that the patient screened more than once was in fact the same patient. See Bailar and MacMahon (1997).

What we have, then, is a hierarchy, with baseline comparisons at the top of the totem pole, more intricate analyses of baseline covariates just below them, followed by analyses of the RPS, followed by automated screening log analyses, and finally scrutiny of the screening log for deferred enrollment at the base. The set of at least apparently reasonable actions would then seem to include the six strategies listed in Table 6.7.

Table 6.7 Apparently reasonable actions regarding testing for selection bias

1. Do not test for selection bias at all.
2. Test for selection bias, but do so using only baseline comparisons across treatment groups.
3. Test for selection bias, but do so using only analyses of baseline variables.
4. Test for selection bias using analyses of both baseline variables and the RPS.
5. Test for selection bias using analyses of baseline variables, the RPS, and the screening log, but only those analyses that can be automated.
6. Test for selection bias in the most comprehensive way possible, using all analyses of baseline variables, the RPS, and the screening log.

One can also 'Consider intermediate, adaptable policy options. Adaptable policy options include a process for obtaining more information, thus reducing uncertainty, and building in decision points to reconsider initial policies' (Stoto, 2002). We first consider the six prospectively defined (non-adaptive) strategies in Table 6.7. Strategy 1 has several clear benefits. For one, it is often the norm and the precedent. Using strategy 1 would require no change to the current 'if it ain't broke, don't fix it' practice. In addition, strategy 1 requires the least amount of work. Finally, strategy 1 will minimize the potential for embarrassing revelations, and will minimize the amount of explanation that will be required. Clearly, strategy 1 is the strategy of choice for any investigator who would choose to engage in selection bias.

Discussing methodological flaws in general, and not necessarily selection bias, Penston (2003, page 119) noted that 'Government, regulatory authorities and health service planners, university professors and researchers, as well as the pharmaceutical industry, have every reason to encourage the current allegiance to the large-scale randomized trial. Whether to provide a foundation for reforms, to preserve reputations and protect personal financial gain, or to secure future profits, maintenance of the *status quo* with respect to megatrials is mandatory. Too much has been invested in this methodology for questions about its validity to be given even so much as a fleeting acknowledgement. The ease with which such questions may be dismissed, however, speaks volumes for the poverty of much of what passes for medical research.'

It is clear that strategy 1 fits in nicely with any push for retaining the *status quo*, instead of digging a little deeper and risking finding something unwelcome. Strategy 1 can also be 'justified' logically, although the logic that justifies it is flawed. As Greenland (1998) points out, the fallacy of affirming the consequent 'embodies an all-too-common approach to 'scientific' inference: A researcher will note that a hypothesis H (often his or her favorite) implies a prediction B, observe that B is indeed what has been observed, and conclude that H must be correct.' The argument that appears to justify strategy 1 is as follow:

Premise 1   No selection bias implies that there is no evidence of selection bias.

Premise 2   To date, there has been no evidence of selection bias.

Conclusion   There is no selection bias.

Of course, this 'logical' argument actually constitutes a fallacy, because the phrase 'there is no evidence that . . .' actually 'encompasses two totally different situations. The first context (and overwhelmingly the commonest) is that there is no evidence because no one has looked. The second is that someone has done the relevant study and found a negative result' (Barraclough, 2003). The only way to distinguish between the two possibilities is to find as severe a test as possible. Understanding that 'theories are provisionally accepted so long as they have not been falsified' (Runde, 1996), one is then not even entitled to *provisionally* believe that there is no selection bias unless one looks for it, and uses a severe test in the process. In other words, there is a problem of non-identifiability, because different values of unknown parameters would give rise to the same data distributions (Greenland and Robins, 1986). In this case, one would expect to see the same apparent superiority of the active treatment group under either of two scenarios, either true superiority of the active treatment group or selection bias.

Recognizing this, strategy 1 might not be viewed so favorably, especially from the perspective of society or public health. If there were no need to evaluate medical interventions as reliably as possible, then there likely would be no randomized trials or evidence-based medicine. We take, then, as a premise the societal need to evaluate medical interventions as reliably as possible. Given this need for reliable results of medical studies, it is clear that strategy 1 is dominated by strategy 2. This is because third-order residual selection bias can inflate both the magnitudes of apparent treatment effects and the likelihood with which apparently significant yet misleading information is found. Note that this tendency transcends the divide among the frequentist, Bayesian, and likelihood philosophies. Third-order residual selection bias can systematically make the sampling distribution of the between-group p-values stochastically smaller (Proschan, 1994), and this is true for all response endpoints.

Third-order residual selection bias can also systematically make the sampling distribution of the posterior probability more favorable to the preferred treatment, and it can also make the likelihood ratio in favor of the preferred treatment stochastically larger. That is, third-order residual selection bias infects not only certain data summaries, but also the very data that serve as the source of all data summaries. As such, all data summaries that do not specifically build in robustness to third-order residual selection bias retain the distortions that occurred upstream, in the data themselves. If treatment decisions are based on trial results and meta-analyses which are distorted, then these treatment decisions will be less than ideal. This can lead to unnecessary morbidity and mortality (Bailar, 1976; Berger *et al.*, 2002).

The societal need to ensure that trial results are as reliable as possible indicates the need to manage third-order selection bias effectively. Strategy 1 makes no serious attempt to accomplish this, but seeks instead to deny the existence of third-order residual selection bias as its own justification. The circularity in this logic is apparent:

1. Do not test for selection bias because it is not a problem.
2. Selection bias is not a problem because we are not aware of it.
3. We remain ignorant of selection bias by not looking for it, and also by denying others access to the data that would allow them to test for it.

If we agree that strategy 2 is preferable to strategy 1, then we are arguing contrary to the established principle that baseline testing should not be performed in randomized trials. Yet, while the arguments against baseline testing in the randomized trial context (Senn, 1994) are logical and compelling, they do not appear to account for the possibility of third-order residual selection bias. Given the ease with which strategy 2 can be implemented, it does seem to be preferable to strategy 1. Now strategy 3 (including comparing the expected predictive ability of each covariate to its subsequent imbalance, as outlined at the end of Section 6.1) is also easy to implement, and is better able to detect third-order residual selection bias than strategy 2 is. Hence, strategy 3 appears to be better than either of its predecessors.

Strategy 3 is an effective method for detecting observable third-order residual selection bias, but it is totally ineffective for detecting unobservable third-order residual selection bias (that is, when the selection covariate is latent and is not substantially associated with any of the measured covariates). Again, bearing in mind the societal need to ensure the validity of trial results, at least as well as possible, a method that simultaneously protects against observable and unobservable third-order residual selection bias would be preferable to a method that protects against just one of these. Strategy 4 is the simplest strategy that simultaneously protects against observable and unobservable third-order residual selection bias. Moreover, strategy 4 can be automated, and does not require the collection of any data not already routinely collected. The cost involved in implementing strategy 4 is a one-time expenditure of a modest amount of programming effort. Once the programs are written to automate the use of strategy 4, the only incremental cost involved in using it for a new trial is the cost in running the program that has already been developed.

Strategy 5 cannot be implemented in the absence of the screening log, so strategy 4 is the best one with the data routinely collected in a randomized trial. However, if the screening log is available, then there appears to be little reason not to use it, at least to implement strategy 5. Strategy 6 is more time-consuming, so there could be legitimate objection to its routine use. Among the prospective strategies, then, it would seem that strategy 4 is ideal in the absence of the screening log, and strategy 5 is ideal in the presence of the screening log. In either case, then, baseline covariates and the RPS should be analyzed thoroughly. We suggest listing in the protocol the baseline variables to be tested for imbalance, specifying one as the 'primary' covariate (among those not used for stratification), and formally testing each baseline variable for imbalance. Associations of covariates and the RPS should be presented, along with the ability of the RPS to predict response within each treatment group (the Berger–Exner test).

We now return to adaptive strategies (such as the decision to record the screening log). Again, with no screening log, strategy 4 is as far as one can go, but with a screening log, one can move beyond strategy 5 to strategy 6. This step may or may not be warranted, but it is certainly reasonable to consider the results of the other analyses before committing to performing the manual search of the screening log for deferred enrollment. So the recommended strategy in testing for selection bias is to use strategy 4 when there is no screening log, and to use strategy 5, or possibly strategy 6 (depending on the results of the other analyses), if there is a screening log. One may still question if there is enough information up front to make the decision to record the screening log. In a perfect world, the answer would be an unequivocal yes. However, recording the screening log does introduce additional costs that may or may not be justified. Consequently, we do not take a position on this issue.

6.10   WHO SHOULD TEST FOR SELECTION BIAS?

Even if it is agreed that randomized trials should be routinely tested for third-order selection bias, one may still ask whether the burden for doing so falls on the sponsor. A partial answer to this question is afforded by the recognition that the sponsor has the most to gain from positive results (in terms of efficacy) of the trial. Another partial answer is afforded by the recognition that the sponsor owns the data, and is therefore in the best position to conduct the analyses that would clarify if there was, in fact, third-order selection bias. Of course, the sponsor is also the party with the most to lose in case what appear to be positive findings are overturned by the demonstration of third-order selection bias. In another context, Rosner and Markowitz (2002) pointed out that 'in the face of data that could prove damaging to the future of the industry, corporations actually sought to deny access to information that public authorities needed to adopt prudent policies'. Indeed,

often, the source of facts and information are the very industries or interests who oppose action. We have seen this with tobacco, lead paint, petroleum, pharmaceutical, and asbestos industries, whose control of information, doctoring of studies, support for biased research, and suppression of information have made it impossible for the public and independent analysts to share in unbiased information. (Kurland, 2002)

It is one thing for the sponsor of a trial to agree in principle that tests of selection bias should be performed, but to prefer not to be the one to perform these tests. It is quite another thing not only to refuse to perform these tests but also to withhold the data necessary to perform them, so that nobody else can perform them either. In the absence of complete data sharing, the sponsor is not only the party in the best position to test for selection bias, but it is also the party in the *unique* position to test for selection bias. Once the data cease to be proprietary, there appears to be little justification for not only refusing to perform essential analyses but also withholding the data that would allow others to perform these analyses. Indeed,

Medical research, even if it is conducted by the pharmaceutical industry, is not solely a commercial enterprise designed to maximize personal gain or company profits. The responsible conduct of medical research involves a social duty and a moral responsibility that transcends quarterly business plans or the changing of chief executive officers. (Psaty and Rennie, 2003)

7

Adjusting for Selection Bias in Randomized Trials

Through the first six chapters, we have discussed how selection bias occurs in randomized trials, evidence that it actually does occur, the effect it has, how to prevent it, and how to detect it when it is there. We did not, however, address what to do about it once it is found. The simplest approach would probably be to simply discard the study in which selection bias is found (or suspected), and start over again. If this approach were satisfactory in the real world, then there would be no need for this seventh chapter. But in the real world, randomized trials consume a great deal of resources, including money, time, and effort. Those who design, execute, analyze, report, regulate, and publish randomized trials would tend to resist having to start over from the beginning once this effort has been mounted.

Consider, for example, the Canadian National Breast Cancer Screening Study (CNBSS) was a randomized trial that has been severely criticized and equally vigorously defended. One of the primary bases for the criticism is the baseline imbalances that were observed, and the lack of allocation concealment. Putting the two together, we see that there is suspicion of selection bias. See Section 3.3.9 for more details regarding these allegations. The importance of this trial for this chapter is the fact that its conclusions have not been dismissed. There is still an effort to salvage information on screening for breast cancer from this study, despite what some would say may be severe biases. How, then, to correct for severe baseline imbalances in a randomized trial?

If ignoring the study is one extreme, then the other extreme would be to simply proceed with the usual analyses, and ignore not the

Selection Bias and Covariate Imbalances in Randomized Clinical Trials   V. W. Berger
© 2005 John Wiley & Sons, Ltd.

study but rather the baseline imbalances. This seems to be the approach suggested by Sleight, Pogue, and Yusuf (2002) in their response to criticism (Taylor, 2002) of the baseline imbalances in the Heart Outcomes Prevention Evaluation (HOPE) Study (Sleight *et al.*, 2001). Specifically, Sleight, Pogue, and Yusuf (2002) pointed out that "None of these selected imbalances was significant [at what significance level was not specified]. Randomisation balances known or measured risk factors as well as unmeasured or unknown factors." As we have seen, randomization does not, in fact, ensure any such thing, and it may be for this reason that baseline imbalances that trend in the same direction cause some researchers to challenge the validity of the findings. See Sections 3.3.2, 3.3.29, and 3.3.30, for example.

Several methods have been proposed for comparing treatment groups in randomized trials with baseline imbalances, and these methods have tended to assume, at least implicitly, that the baseline imbalances were random. That is, no effort was made to determine if the imbalances could have been the result of selection bias, or to adjust for it accordingly. See, for example, Wei and Zhang (2001). Rothman (1977) suggests that 'What matters in a particular trial is whether confounding in this trial is present and, if so, to what extent. It is of no interest to learn whether the confounding which exists might be compatible with chance as to its etiology.' If this is true, then any method appropriate for dealing with random baseline imbalances will also be appropriate for dealing with selection bias. But the problem with these adjustment methods, at least when applied to trials that may have selection bias instead of (or in addition to) random covariate imbalances, is that they are limited by the covariates that are actually measured (Moses, 1995).

Recall that selection bias may result in an unbalanced unmeasured covariate, but no observable baseline imbalances (Berger and Exner, 1999). As Mitchell (1981) noted in the context of the Norwegian Timolol Trial (which had numerous baseline imbalances; see Section 3.3.27), "The authors suggest that by 'adjusting for the largest differences and other factors considered prognostically important' they can overcome the inequality of the two groups, but I do not believe that this can be done. The techniques used can only adjust for overt variables, and if, as a group, the placebo patients also have related but hidden adverse factors such as platelet hyperreactivity then no amount of adjustment for the overt differences can overcome this hidden initial imbalance."

The goal of this chapter, then, is to provide rigorous methodology for comparing treatment groups in the presence of selection bias in randomized trials. In Section 7.1 we first review some methods that have been proposed in the literature for dealing with baseline imbalances when it was suspected that the nature of these imbalances was not random. In Section 7.2 we discuss a complete lack of allocation concealment. In Section 7.3 we discuss imperfact allocation concealment as when upcoming allocations can be predicted based on past ones, but cannot be directly observed. This latter case is much more amenable to treatment than is the former case. In particular, we discuss appropriate approaches based on the reverse propersity score or RPS in Section 7.3.1, and a likelihood-based approach in Section 7.3.2.

7.1 METHODS PROPOSED FOR ADDRESSING NON-RANDOM BASELINE IMBALANCES

As discussed in Section 3.3.27, Mitchell (1981) tried to quantify the p-value for mortality by conditioning on the most extreme baseline p-value. In particular, the method was to divide the efficacy p-value (0.001) by the most extreme baseline p-value (0.01) to obtain 0.1. As pointed out in Section 3.3.27, this method fails to account for the multiplicity of baseline imbalances, and hence is rather crude. Tarone (1995) suggested eliminating from the analyses of the Canadian National Breast Screening Study (CNBSS) all advanced cases detected by physical examination at the initial screening visit. Gotzsche and Olsen (2000) suggested a similar approach, specifically identifying those mammography subjects who had been randomized properly, presumably so that the subjects who were not randomized properly would be discarded.

Along the same lines, Peto (1999) stated that "it would be useful to try to recover a properly randomized comparison from the CAPPP study. Perhaps it could still, at this late stage, be determined which centers sometimes broke the rules in this way, yielding inappropriate foreknowledge of the next treatment ... If this can be done reliably, then a report on outcome by allocated treatment should be published that includes only those centers where randomization can be trusted not to have been distorted by foreknowledge of the next treatment allocation. Alternatively, maybe some other way can be found (e.g., by exclusion of certain time periods, or of certain time periods at certain centers) to publish results only from patients who are known to have been properly randomized." The general approach may be described as eliminating from the analysis any subjects with a certain characteristic that happens to be unbalanced at baseline across the treatment groups. But this elimination approach has problems.

In addition to constituting a deviation from the intent-to-treat (ITT) approach (Fergusson *et al.*, 2002) and the loss of information, this approach also ignores the fact that an imbalance with respect to the presence of a certain characteristic would generally imply an imbalance also with respect to the absence of that characteristic. So it may not be clear if it is those subjects possessing or lacking the characteristic that should be eliminated. And what would one do with a covariate that were not binary, but rather continuous or ordered categorical? One could also compare the treatment groups separately for those with and without the characteristic in question, or stratify by the unbalanced covariate. This would be the logical extension of the elimination approach, and it applies also when the covariate is categorical. However, we are then back to the inherent limitation that some key covariates will generally not be measured.

When a comprehensive cohort follow-up study (Olschewski *et al.*, 1992) is used, there would be data available on all screened patients, even those who were not randomized. This would allow for an "intent-to-randomize" approach to analysis. Specifically, one could try to determine, using whatever patterns of selection bias that might have been found, which patients would have been randomized, and to which groups, had a different allocation sequence been observed. This can be used to construct an appropriate permutation test, as described by Berger and Exner (1999), in their Point #7 (page 325). One could also directly include excluded patients as if they had been randomized to the group to which they would have been randomized. See Section 2.4 of Swingler and Zwarenstein (2000) for the description of an example of this approach. More often, however, data are available only for those patients who are actually randomized, so we consider only this situation in the remainder of the chapter.

7.2 SELECTION BIAS ARISING FROM A COMPLETE LACK OF ALLOCATION CONCEALMENT

Since there was no allocation concealment in the CNBSS, and every future allocation was observable and known with certainty, the selection bias that would have been possible in this trial would have been different from the selection bias that is possible when there is at least an attempt at allocation concealment. That is, the quantity $P\{E\}$ discussed in Chapter 2, which is essentially the probability of receiving the active treatment given the prior allocations and the restrictions on the randomization, is identical to the actual assignment to E (is 1 when E is assigned, and 0 when E is not assigned) when there is no allocation concealment. When there is allocation concealment, so that not every future allocation is known with certainty, $P\{E\}$ may still be exploited, but at least it is not identical to the treatment allocated. This difference allows for a separation of effects of E and effects of $P\{E\}$. Such separation of effects is not possible when there is no allocation concealment, as E and $P\{E\}$ are completely confounded. In this case, it may be impossible to salvage useful comparative information, and each treatment group in the study is more like a one-arm (uncontrolled) study into itself. In this case, the results should probably be summarized with descriptive statistics, but not with inferential statistics such as p-values.

7.3 SELECTION BIAS ARISING FROM IMPERFECT ALLOCATION CONCEALMENT

If there is allocation concealment, as the term is usually defined (that is, the process), then it is not the case that every future allocation can be observed. While allocation concealment can be subverted in more direct ways (such as by holding sealed envelopes up to a light), we consider only indirect subversion of the type we defined in Chapter 2, specifically predicting (rather than observing) future allocations based on the known restrictions on the randomization and the known prior allocations. For simplicity, then, we will consider unmasked trials, although we note that this material is relevant to masked trials as well, at least to the extent that their masking is not perfect. When future allocations can be predicted but not observed, it will turn out that only some of these predictions are known ahead of time to be correct.

For example, if prediction is based upon the randomized block design with a fixed block size of 4, then the last allocation in each block will be deterministic, as will the second last allocation in each block of the type AABB or BBAA. In each block, the second allocation will be predictable, but will not be deterministic. This means that P{E} will not be 0.5, and so prediction is possible, but it will not be 0 or 1, so this prediction may be wrong. We can exploit this fact, that the actual treatment assignment may not follow the predicted treatment assignment, to study the effects of each of these. One method for separating the effects of the predicted treatment assignment (i.e., selection bias) from the effects of the actual treatment effect is to include both in the model used to evaluate efficacy. For example, if the efficacy endpoint is survival, then one would build a prediction model for survival time by including as predictors both the actual treatment assignment and P{E}, the predicted treatment assignment. This value P{E} of the predicted assignment has been called the reverse propensity score (Berger, 2005b) in a paper that further develops the correction method based on this score (RPS). In the present chapter we will consider both this RPS approach and a competing model-based method developed by Ivanova *et al.* (2005).

7.3.1 The RPS Approach to Adjusting for Selection Bias

Recall the Berger–Exner graphs presented in Section 6.5, and how they were used to detect selection bias by comparing response variables across RPS levels within each treatment group. These same Berger–Exner graphs are the key to using the RPS to adjust for selection bias. To see this, consider a rather extreme Berger–Exner graph, based on the fictitious data in Table 7.1 (also Table 2 of Berger, 2005), which is reproduced in this section.

Suppose that the covariate is a perfect predictor of response, so that each patient with covariate value 1 responds, and no patients with covariate value 0 respond. Then the baseline imbalance would create

Table 7.1 Associations among a binary covariate, the RPS, and the treatment group. The reverse propersity score to detect selection bias and correct for baseline imbalances, Berger V. W., 2005, copyright John Wiley & Sons, Ltd. reproduced with permission.

	RPS = 0.0	RPS = 0.5	RPS = 1.0	Total
Control	0/10	5/10	0/0	5/20
Active	0/0	5/10	10/10	15/20
Total	0/10	10/20	10/10	20/40

the illusion of a treatment effect, because the response rate would be 15/20 (75%) in the active group and only 5/20 (25%) in the control group. The Berger–Exner graph would be as follows:

$Y = 1$		0000011111	1111111111
$Y = 0$	0000000000	0000011111	
	RPS = 0.0	RPS = 0.5	RPS = 1.0

Figure 7.2.1.1 The Berger–Exner Graph for the Fictitious Data in Table 7.1.

The comparison of treatment groups when using the RPS to adjust for selection bias occurs only within levels of the RPS. One drawback to this approach is the loss of information, as no comparisons exist at the extreme RPS values of 0 and 1. When the block size is two, as it is for the fictitious data in Table 7.1, this means that the only comparison occurs for the patients randomized when the RPS value was 0.5. Of course, with larger block sizes, there would be more RPS values that allow for treatment comparisons, and so correspondingly fewer patients would be lost in the analysis. Anyway, returning to the fictitious data in Table 7.1, we see that when the RPS value is 0.5, each treatment group has five responders and five non-responders, for a 50% response rate. The RPS has clarified that the apparent treatment effect was due to nothing more than selection bias, and that when the selection bias is removed, there is no treatment effect whatsoever. We could have created a data set in which the direction of the effect would be reversed, as well. Consider, for example, Table 7.2.

   The overall response rates are now 60% (12/20) in the active group and 40% (8/20) in the control group, but when using the RPS as a

Table 7.2   Associations among a response variable, the RPS, and the treatment group

	RPS = 0.0	RPS = 0.5	RPS = 1.0	Total
Control	0/10	8/10	0/0	8/20 (40%)
Active	0/0	2/10	10/10	12/20 (60%)
Total	0/10	10/20	10/10	20/40 (50%)

predictor, either with the Berger–Exner graph or with the table structure in Table 7.2, it is clear that the control is more efficacious, as it has an 80% (8/10) response rate among unselected patients (when the RPS value is 0.5), as compared to only 20% (2/10) for the active group. The RPS can be used as a predictor with any type of response variable. If the response variable is close to normally distributed, then one can use the analysis of covariance (ANCOVA) model. With a time-to-event endpoint, one can use Cox proportional hazards regression. Mantel-Haenszel methods can be used for categorical data. The RPS can be used in place of other covariates or along with them.

If it turns out that the direction of the effect varies across RPS values – that is, each treatment group appears to be substantially better for some RPS values – then this would be a rather curious situation, and would not fit into the classical paradigm of what one would expect from selection bias. In such a case, the comparison at the "neutral" RPS value is probably the most important one. In two-arm trials with 1:1 allocation, the neutral RPS value is 0.5, or the unconditional probability of allocation to the active treatment group. With three groups and 1:1:1 allocation, the neutral value is 0.33. Clearly, this value changes with unequal allocation. For example, it would be 0.67 with 2:1 allocation in favor of the active treatment group. The reason for favoring the comparison at this RPS value is that it is, in some sense, the most pure, and free from selection bias. Even with selection bias, the investigators have no incentive to recruit extreme patients (either especially likely to respond or especially unlikely to respond) at the neutral value of the RPS, and so this RPS value might offer the most external validity.

7.3.2 The Ivanova–Barrier–Berger (IBB) Method for Correcting Selection Bias

Consider a two-arm unmasked trial with experimental (E) and control (C) treatments using the randomized block design with fixed, known block size, and consider also a binary outcome. Then at each stage of the patient recruitment process the investigator knows P{E}, the probability that the experimental treatment will be assigned to the next enrolled patient. Following Ivanova *et al.* (2005), let the treatment indicator variable X_i, $i = 1, 2, \ldots, n$, equal 1 if the ith patient is assigned to treatment E, or 0 if the patient is assigned to C. Each patient has one potential outcome if E is received and another if C is received. Suppose that candidate patients can be classified at baseline as strong (success probability $p_1 + \delta$ or $p_2 + \delta$, respectively, for E and C); medium (success probability p_1 or p_2); or weak (success probability $p_1 - \delta$ or $p_2 - \delta$), where the parameter δ must satisfy $0 \leq \delta \leq \min(p_1, p_2, 1 - p_1, 1 - p_2)$. This additive model is probably simpler than a more realistic one in which the covariate parameter δ is added not to the success probability itself but rather to the log-odds $\log(p/(1-p))$, to obtain a new log-odds which can then be inverted to induce a new success probability.

We proceed with the simplistic additive model (on the original scale), and, following Berger *et al.* (2003a), assume that the investigator will enroll a strong patient if P{E} exceeds a fixed cutoff G or a weak patient if P{C}= 1−P{E} exceeds the same cutoff. Otherwise, a medium patient is enrolled. Note that this is a different model from the one discussed so far, in which unselected patients were to be enrolled when P{E} = 0.5. That is, up until this point we have considered the situation in which strong, weak, or medium patients could be enrolled when P{E} = 0.5. Now we consider instead the different situation in which only medium patients are to be enrolled when P{E} = 0.5. We assume that there are enough strong, medium, and weak patients in the population, so that a patient of each kind is always available if needed. We define the patient response type as $S_i = 1.0$ for a strong ith patient, $S_i = 0.5$ for a medium ith patient, and $S_i = 0.0$ for a weak ith patient. Also, the binomial outcome Y_i takes the value 1 or 0 for success or failure. Ivanova *et al.* (2005) derive the likelihood function by noting that

$$
\begin{aligned}
E\,(Y_i\,|\,X_i,\,S_i) &= X_i\,p_1 + (1 - X_i)\,p_2 + S_i\delta - (1 - S_i)\,\delta \\
&= X_i\,p_1 + (1 - X_i)\,p_2 + \delta\,(2S_i - 1) \\
&= (p_2 - \delta) + (p_1 - p_2)\,X_i + 2\delta S_i.
\end{aligned}
$$

This means that the likelihood function is $L = \prod_{i=1}^{n} m_i^{y_i}\,(1 - m_i)^{1 - y_i}$, where n is the total number of patients in the trial (in previous chapters we used $2n$ for this quantity), and $m_i = E\,(Y_i\,|\,X_i,\,S_i)$ as defined above. This likelihood can be maximized, subject to the necessary constraints $0 \le p_1 \le 1$, $0 \le p_2 \le 1$, and $0 \le |\delta| <$ $\min\,(p_1,\,p_2,\,1 - p_1,\,1 - p_2)$, to allow for the estimation of parameters and the testing of hypotheses. The 'strength' or 'weakness' or 'mediocrity' of any given patient is a baseline covariate that is not recorded (and may or may not be observed in any given trial, but we assume that it is). Generally, when one thinks of a baseline covariate, one thinks of the patient bringing that to the merging of patient and accession number. But in this case, since a patient with a given level of strength will be sought based on the cutoff G (which is known to the investigator but is not generally revealed to any other party) and P{E}, which depends on the accession number but not the patient, the situation is reversed. That is, it is the accession number that determines the 'covariate' and, indirectly, the identity of the patient (the first qualified patient screened having this attribute). Hence, S_i is quite a peculiar covariate.

We return now to the problem of maximizing the likelihood, subject to the necessary constraints. Because the cutoff value G is not known to the analyst, it must be estimated. One way to do so is as follows. For most reasonable block sizes, there are very few cutoff values that need to be considered, because any two cutoff values that both fall between the same pair of consecutive attainable values of P{E} are equated by this model. That is, if the block size is 2, then P{E} can assume only the values 0.0, 0.5, and 1.0. This means that the only relevant values for the cutoff are less than 1.0 and 1.0 (in which case there is no selection bias) and less than 1.0 (in which case there is). If the block size is 4, then P{E} can assume only the values 0.00, 0.33, 0.50, 0.67 and 1.00 (see Section 4.3). Again, the cutoff matters only to the extent that it indicates which of these intervals, [0.5, 0.67), or [0.67, 1.00) it falls in (or it can be equal to 1.00 if there is no selection bias). There are only three possibilities. Given the block size, and the number of choices for the cutoff it induces, we maximize the likelihood separately for each admissible value of the cutoff, and then select that value that leads to the maximum of all of these maximums.

An alternative to the Berger–Exner test of selection bias (Berger and Exner, 1999; also discussed in Section 6.5 of this book) can be based on testing the hypotheses H_0: $\delta = 0$ versus H_A: $\delta \neq 0$. In this chapter we are more concerned with testing for a difference in treatment effects given the presence of selection bias. This test can be based on testing the hypotheses H_0: $p_1 - p_2 = 0$ versus H_A: $p_1 - p_2 \neq 0$, where now δ is considered a nuisance parameter. As Ivanova *et al.* (2005) point out, if $\delta = 0$, then there is no cutoff point to estimate and the model loses one parameter. These authors approximated the reference distribution using a chi-square distribution with one degree of freedom for use in a simulation study with randomized blocks with fixed block size 6.

The simulations are based on 5000 runs, and each used a sample size of 192, with a nominal significance level of $\alpha = 0.05$. The cutoffs were 0.5 and 0.99, and were estimated correctly in about 75% of runs. The more important question, for our purposes, is whether the model can be used to adjust the treatment comparison for selection bias. Table 7.3 of Ivanova *et al.* (2005) reveals that the true Type I error rate can be grossly increased by selection bias. For example, if the null hypothesis is true ($p_1 - p_2 = 0$, both 0.50), but $\delta = 0.20$ and the cutoff is 0.50, then a nominal 0.05-level test will have an actual level of 0.54, which is more than 10 times as large as it should be.

The true test of this new method for correcting for selection bias is in the true Type I error rate of this procedure in this situation, and in similar situations. For this particular situation, the true Type I error rate is 0.04, which is close to the nominal 0.05. Changing the cutoff to 0.99 but keeping $p_1 - p_2 = 0$ and $\delta = 0.20$, the true Type I error rate is 0.29 if selection bias is ignored, and 0.05 (exactly right) if selection bias is addressed with this new method. Furthermore, if the control treatment is superior to the active treatment, then this true effect may be offset by selection bias so that no effect is found. For example, if $\delta = 0.20$ and the cutoff is 0.50, and the response rates are 0.50 and 0.70 in the active and control groups, respectively, then the power to detect this true difference is only 0.11 if selection bias is ignored, but 0.78 if selection bias is addressed with this new method.

Table 7.3   Power and type I error rates for the test of H_0: $p_1 = p_2$ using linear model (2) and logistic regression model (4). Adjusting for observable selection bias in block randomized trials, Ivanova, A., Barrier, R. C. and Berger, V. W. 2005, copyright John Wiley & Sons, Ltd. Reproduced with permission

			Cut-off = 0.50			Cut-off = 0.99		
p_1	p_2	δ	Linear model	Logistic model	Unadjusted	Linear model	Logistics model	Unadjusted
0.50	**0.50**	**0.00**	0.06	0.05	0.05	0.06	0.05	0.05
0.50	**0.50**	**0.20**	0.04	0.05	0.54	0.05	0.04	0.29
0.50	**0.70**	**0.00**	0.65	0.64	0.80	0.65	0.63	0.80
0.50	**0.70**	**0.10**	0.73	0.64	0.43	0.68	0.67	0.56
0.50	**0.70**	**0.20**	0.78	0.66	0.11	0.70	0.67	0.29
0.60	**0.40**	**0.00**	0.63	0.67	0.79	0.63	0.66	0.79
0.60	**0.40**	**0.30**	0.81	0.67	>0.99	0.67	0.64	>0.99
0.70	**0.50**	**0.00**	0.65	0.69	0.81	0.65	0.67	0.81
0.70	**0.50**	**0.10**	0.68	0.69	0.98	0.66	0.67	0.95
0.70	**0.50**	**0.20**	0.76	0.66	>0.99	0.68	0.65	0.99
0.70	**0.70**	**0.00**	0.06	0.04	0.05	0.06	0.04	0.05
0.70	**0.70**	**0.20**	0.04	0.04	0.62	0.04	0.05	0.34

The data are generated according to model (2).

In their Table 7.4, Ivanova *et al.* (2005) also considered the more realistic scenario in which the investigator may *attempt* to enroll a strong patient or a weak patient but he or she is able to do so only 50% of the time, because strong and weak patients are not always available. Only medium patients are always available. The results here as equally impressive as with the previously considered case in which an investigator can recruit any type of patient all the time. That is, the power is still large when there is a true effect to detect, and the true Type I error rate is still close to the nominal Type I error rate. What has changed is that now the true Type I error rate of the usual procedure (ignoring selection bias) does not inflate the true Type I error rate by as much as it did in the previous case. For example, if the null hypothesis is true ($p_1 - p_2 = 0$, both 0.50), but $\delta = 0.20$ and the cutoff is 0.50, then a nominal 0.05-level test will have an actual level of 0.17. If the cutoff is 0.99, then a nominal 0.05-level test will have an actual level of 0.11.

It seems reasonable to always test for selection bias, and then to use one of the adjustment techniques discussed in this chapter (either the

Table 7.4 Power and type I error rates for the test of $H_0: p_1 = p_2$ using linear model (2) and logistic regression model (4). Adjusting for observable selection bias in block randomized trials, Ivanova, A., Barrier, R. C. and Berger, V. W. 2005, copyright John Wiley & Sons, Ltd. Reproduced with permission

			Cut-off $= 0.50$			Cut-off $= 0.99$		
p_1	p_2	δ	Linear model	Logistic model	Unadjusted	Linear model	Logistics model	Unadjusted
0.50	**0.50**	**0.20**	0.05	0.05	0.17	0.06	0.05	0.11
0.50	**0.70**	**0.02**	0.73	0.67	0.43	0.68	0.65	0.56
0.60	**0.40**	**0.30**	0.69	0.63	0.99	0.69	0.63	0.97
0.70	**0.50**	**0.20**	0.69	0.67	0.98	0.66	0.65	0.94

The data are generated according to model of enrolment with limited availability. The sample size of 192 is used in each trial.

reverse propensity score or the likelihood-based method, or both) if the likelihood of selection bias appears to be substantial. The evaluation of the likelihood-based method was restricted to unmasked trials, in which all prior treatment assignments are known. In masked trials some prior allocations would likely remain unknown, but some may be known, as well, and so these issues do not disappear. There may still be value in adjusting for selection bias even in masked trials, although the benefit of doing so may be less than the benefit of doing so in unmasked trials.

8

Managing Selection Bias in Randomized Trials

The first seven chapters of this book provided information on how selection bias occurs in randomized trials and how to prevent, detect, and correct for it. The purpose of this chapter is to put all this material together to provide a set of items that can be used to collectively manage selection bias in randomized trials. Not every issue has a clear-cut solution, however, and so there are also open questions, which will hopefully lead to further research, and, ultimately, recommendations that are not controversial.

8.1 ACTION POINTS DURING THE DESIGN PHASE OF THE TRIAL

During the design phase of the trial, many decisions need to be reached. These include the patient population, the allocation procedure, the decision to mask or not, the decision to conceal the future allocations or not, and the choice of variables to be captured on the case report forms. Each of these issues has implications for the susceptibility of the trial to selection bias, as we have seen in the previous chapters. If either masking or allocation concealment can be achieved perfectly, then there can be no selection bias. The problem is that, in general, neither one can be achieved without sacrificing something else (for example, using unrestricted randomization could allow for successful allocation concealment, but would also enable chronological bias).

Selection Bias and Covariate Imbalances in Randomized Clinical Trials   V. W. Berger
© 2005 John Wiley & Sons, Ltd.

The solution appears to be to still go through the processes of masking and allocation concealment (that is, no treatment identities are intentionally revealed until after the trial is complete and the database is locked), and to comment on how successful these steps are likely to be. The patient population is a trickier issue, because it is dictated by concerns having nothing to do with selection bias. Still, if selection bias is to be considered, then what are the implications? On the one hand, selection bias requires that the investigator be able to turn patients away when their set of covariates makes them a bad match for the up-coming treatment. In a large simple trial (Peto *et al.*, 1995), one salient feature of which is the minimal set of entry criteria, it may be harder to deny enrollment to qualified patients than it would be otherwise (with more entry criteria), so one can see how a broad patient population might minimize selection bias. But, on the other hand, selection bias also requires that some patients appear (at baseline) to be better po-tential responders than others, because if a study were performed in a homogeneous group, or even in clones, then there would be no oppor-tunity to differentially recruit patients for the treatment groups. This means that perhaps a narrowly defined patient population might be best for controlling selection bias. More homogeneous patient groups would also allow for greater separation between treatment groups, so that hopefully an ideal treatment could be identified for each particu-lar patient profile, even if different profiles require different treatments (Penston, 2003, page 130).

We will consider the issue of the 'right' patient population, broad or narrow, to be an open question, and one that might have different valid solutions in different contexts. Likewise, we consider the choice of the maximal tolerated imbalance to be at the discretion of the research team, and again, there might be different valid solutions in different contexts. However, one point deserves to be made in the case in which multiple trials are considered for a given treatment. In such a case, one would not want to select the same maximum tolerated imbalance, or the same patient population, for both (or all) of the trials. Varying these parameters will vary also the susceptibilities of the trials to the various biases (for example, selection bias and chronological bias), but keeping them the same will not.

Rosenbaum (2001) points out that

an effective research design would replicate the treatment without replicating whatever hidden biases were originally present . . . The mere reappearance of an association does not convince us that the association is causal – whatever produced the association before has produced it again. It is the tenacity of the association – its ability to resist determined, planned, deliberate challenges – that ultimately convinces us . . . A replication is trivial if it reproduces the original study exactly, increasing the sample size, decreasing sampling variability, but shedding no light on hidden biases . . . Absent efforts to prevent replication of hidden bias, there is little hope that mere repetition eliminates bias. Biases occur for reasons.

It would seem to be best, then, to vary as much as possible across trials, in accord with the principle of variative (or Baconian) induction, the fundamental feature of which "is that observations are performed under different conditions which bring to light factors which potentially challenge the reliability of a generalization" (Penston, 2003, page 26).

Regarding the choice of baseline covariates to collect, there are again pros and cons to collecting more, rather than less. On the one hand, if every conceivable baseline covariate is collected, and there is selection bias, then the selection covariate (used to discriminate strong patients, suitable for one treatment group, from weak patients, suitable for the other treatment group) would be observed. This means that any selection bias would have to be observable selection bias, and simple adjustment for the unbalanced covariate(s) might suffice as a solution. On the other hand, the greater the number of covariates collected, the better the investigator is able to distinguish strong patients from weak patients, so this might also enable selection bias. We consider this to be another open question.

Of course, not all covariates are created equal. Some are much more prognostic than others. The more prognostic a covariate happens to be, the more important it is that it be balanced across treatment groups. For this reason, it is a good idea to do more than simply list the covariates to be considered. It would be helpful to also summarize any previous studies that might inform the consideration of how prognostic each covariate is. It might also be a good idea to rank the covariates based on how prognostic they are expected to be, based on these previous studies. Such a ranking might help build a model during the analysis stage, and might also come into play if it turns out that some covariates are well balanced and others are not. That is, the ranking would clarify which covariates to be especially concerned about.

Another issue to consider during the design stage is the allocation procedure. In Chapter 5 we discussed various randomization

Table 8.1   Items to consider in the design stage of a randomized trial

---

Always plan to mask the trial in terms of the process of masking.
Discuss openly the extent to which the outcome of masking can be achieved.
Always plan to conceal the future allocations.
Discuss openly the extent to which the outcome of allocation concealment can be achieved.
Carefully consider the patient population.
Carefully consider the set of covariates to measure.
Describe how prognostic each covariate is expected to be, and rank the covariates.
Carefully consider the maximum tolerated imbalance.
Carefully consider if terminal balance is needed.
Decide on the maximal, randomized blocks, or some other randomization procedure.

---

techniques, and these can be considered for any randomized trial. There are reasons why the maximal procedure might be preferred to the randomized block procedure, but even if the maximal procedure is selected (within strata), this still does not completely specify the randomization technique. What should the maximum imbalance be? Should terminal balance be forced? If the maximum tolerated imbalance is not too large, then there is probably little benefit in forcing terminal balance in the size of the treatment groups, but how small the allowed imbalance would have to be is subjective. Table 8.1 displays the recommendations and controversies that arise during the design stage.

8.2 ACTION POINTS DURING THE CONDUCT OF THE TRIAL

Another set of issues arises as the trial is being executed. For example, what (if anything) should be done in case the masking must be broken for a given patient with a severe adverse event? What information (if any) should be recorded for a patient who is screened but not enrolled? If a run-in phase was used prior to randomization, then this question comes in two parts: what information should be collected on those patients denied entry into the run-in phase, and what information should be collected on those patients excluded during (or after) the run-in and prior to randomization? Should the patients who were randomized be asked (after the study is complete and the database is locked) what treatment they thought they received? Should the investigators be asked this question?

Another issue arises when the masking is intentionally broken for a given patient due to an adverse event that appears to require knowledge of the treatment condition for proper management. In fact, often the masking can be retained even in this case without adversely affecting the quality of care for the patient in question (Ayala and MacKillop, 2001), but nevertheless it is not uncommon to break the masking anyway. As Berger and Exner (1999) pointed out, there is a danger in breaking the masking for a given patient if the block to which this patient was randomized is still accruing. That is, consider a masked trial with blocks of size 2. Suppose that the first patient in a block has an injection site reaction, and the investigators need to know the treatment assigned, before the second patient in this block has been identified. In such a case, allocation concealment is compromised for the second patient in this block, so Berger and Exner (1999) proposed closing this block to enrollment and proceeding to the next block. But what if the randomized blocks procedure is not used?

Suppose, for example, that the maximal procedure is used instead. Then there are no blocks, so the proposed step would not apply. But, as Berger *et al.* (2003a) pointed out, it would be nearly impossible to predict future allocations generated by the maximal procedure based on knowing only one or several of the previous allocations. In such a case, then, it may be acceptable to simply proceed with no modification. On the other hand, if the situation merits concern and the need for a remedial action, then it would be possible to consider all allocations up until this time to be a single (possibly very large, depending on the accession number of the patient in question) block, and then to start over with a new maximal procedure. That is, the imbalance would be reset to zero after this patient, thereby making it difficult to predict future allocations, and impossible to do so based on previous allocations.

Regarding information to be recorded for a patient who is screened but not enrolled, Berger and Exner (1999) described methods for correcting selection bias by making use of response data on such patients, and methods for detecting selection bias based on covariate data on such patients. There are reasons, then, to conduct a comprehensive cohort follow-up study (Olschewski *et al.*, 1992) and obtain complete data on even the patients who were not randomized. It is unclear how this would apply to the case in which some patients are not randomized because they were denied entry into the run-in phase, and others are not randomized because they were excluded during (or after) the run-in and prior to randomization. It is known, however, that selecting patients based on their experiences during a run-in leads to a different type of selection bias (Berger *et al.*, 2003b), so this practice should not be used. Of course, the benefit of collecting information on screened patients who are not subsequently enrolled needs to be balanced against the resources it would require.

Another issue is whether or not the patients who were randomized should be asked (after the study is complete and the database is locked) what treatment they thought they received, and if the investigators should be asked this question. One issue is the use to which such data would be put. Another issue is the reliability of the data (Fergusson *et al.*, 2004b). If there is selection bias, then presumably the patients would not be a party to it, so it is hard to imagine patients having a reason to conceal the truth regarding their views on which treatment they believe they received. However, selection bias cannot occur without at least some investigators causing it, and so if there is selection bias, and we ask those investigators who caused it which treatments they think each patient received, then we are essentially asking them to confess. An investigator who thinks this through might be inclined to avoid self-incrimination, and might provide misleading answers to these questions. Such answers might suggest that there was no selection bias, but again, the reliability of the source of these answers needs to be considered.

One step that might convince investigators to provide honest answers to this question would be the use of the randomized response technique (see Singh and Mathur, 2004, for a recent discussion of this technique), which can be modified to fit this situation as follows. For each patient, the investigator tosses a penny and a nickel. If the penny turns up heads, then the investigator answers the question based on the outcome of the nickel (for heads, record that you thought that the active treatment was received; for tails, record that you thought that the control was received). If, however, the penny turns up tails, then you have to answer the question honestly, and report what you thought the patient received. The key here is that the result of the toss of the penny is not recorded, so nobody but the investigator knows which question was being answered. Because the same investigator would be answering many questions (one for each patient he or she randomized), and in the case of selection bias would be concerned with the pattern that might emerge (as opposed to being concerned with any one response), there is reason to believe that this technique would not work as well in this situation as it would in the more usual case in which each respondent answers just one question. Still, it might be expected to work better than directly asking an investigator to record the treatment assignment he or she thought was made to each patient.

The question remains of how to use data on which treatment a patient or investigator thought was received. Because this type of data may be unreliable (as mentioned above), it is quite useful at one end of its spectrum and much less useful at the other end. Specifically, if it is found that many more than the expected number of allocations one could guess by chance alone (50% in a study with two arms and 1:1 allocation, for example) were correctly known, then this suggests that the study was not masked very well. If only the expected number of correct guesses are reported, then on the surface this suggests that the masking was intact, but it may also conceal a reluctance to report the truth. Another distinction between the responses of patients and investigators is that the patients would presumably not know the restrictions on the randomization, and would not know any prior allocations. Their answers would then be expected to be based exclusively on their own experiences.

Conversely, an investigator can form an opinion about a treatment received before it is even administered, based on prior allocations and knowledge of the restrictions on the randomization. That is, the investigator will probably follow the restrictions of the randomization, at least in his or her thinking, if not in his or her reporting. One could study the responses of an investigator (especially if randomized response is not used) to see if these responses follow the pattern mandated by the restrictions on the randomization, and to see, for example, if there are more correct guesses at the end of blocks than at the beginning of blocks. This would suggest the mechanism for guessing. Table 8.2 displays the recommendations and controversies that arise during the conduct of the trial.

Table 8.2   Items to consider during the conduct of a randomized trial

---

Consider if any remedial action is needed in the case of intentional
unmasking of a patient.
Consider recording covariate and response data on even patients not
randomized or enrolled.
Do not use experiences during a run-in to select patients for subsequent
randomization.
Consider asking patients which treatment they thought they received.
Consider asking investigators which treatment they thought they received.
Consider using randomized response to increase the chances of obtaining
honest answers.

8.3   ACTION POINTS DURING THE ANALYSIS OF TRIAL DATA

A final set of issues arises during the analysis of the trial. For example,
would the analysis be modified if a covariate not prospectively specified
in the model is grossly unbalanced? If so, then how? Would it matter
if an observed covariate imbalance appears to be random, as opposed
to the result of selection bias? How would one even be able to tell the
difference? These are a few of the important issues to address during
the analysis stage, and are probably the issues most closely related to
selection bias. We saw in Chapter 7 that there are new techniques
available for correcting a treatment comparison for selection bias, so
in fact selection bias would trigger a different analytic strategy than a
random covariate imbalance would. That is, it is worth knowing if an
observed covariate imbalance is random or the result of selection bias.

How, then, can one distinguish a random covariate imbalance from
selection bias? One way is by using the Berger–Exner test to study the
relationship between the predicted treatment assignment $P\{E\}$ and
the outcome within each treatment group. Related analyses can be
used to study the relationship between $P\{E\}$ and the key covariates.
This analysis is facilitated by the ranking of the covariates mentioned
earlier in this chapter. In the absence of selection bias, $P\{E\}$ should be
related to the outcome measures (to the extent that there is a true treatment effect, this relationship between $P\{E\}$ and the outcome measures
will reflect this true treatment effect), but $P\{E\}$ should not be related
to any outcome measure within a treatment group. If it is, then we

certainly have evidence of selection bias. Likewise, if P{E} is related to any covariate, then this is evidence that there is selection bias, and, further, that this covariate is either related to the selection covariate or is the selection covariate itself. It is also possible to detect selection bias by comparing covariate values of those patients randomized and those patients not randomized by P{E} and perhaps by the treatment group that was assigned or that would have been assigned had the patient been randomized (Berger and Exner, 1999). Both are important because they reveal different mechanisms for selection bias. In particular, if one can predict the randomization decision with the combination of P{E} and a key covariate, then this would suggest both that there was selection bias and that P{E} was used to predict upcoming allocations. On the other hand, if the actual allocation itself, but not P{E}, can be used along with a covariate to predict the randomization decision, then this would suggest instead that the mechanism that enabled selection bias involved direct observation of the upcoming allocations, perhaps by tampering with sealed envelopes. It is helpful to use the flow diagram in figure 2.6.1 to distinguish a random imbalance from selection bias caused by direct observation of upcoming allocations from selection bias caused by prediction of future allocations based on P{E}. Table 8.3 displays various situations, and the importance of conducting the Berger–Exner test in each.

Table 8.3 The importance of conducting the Berger–Exner test

	Fixed block size	Varying block sizes	Blocks not used
Unmasked, imbalances found	Essential	Essential	Essential
Unmasked, no imbalances found	Essential	Essential	Highly desirable
Single-masked, imbalances found	Essential	Highly desirable	Highly desirable
Single-masked, no imbalances found	Essential	Highly desirable	Desirable
Double-masked, imbalances found	Highly desirable	Desirable	Desirable
Double-masked, no imbalances found	Desirable	Desirable	Optional

If selection bias is rigorously sought, and not found, then this suggests that it is safe to proceed with the usual analyses. If, however, selection bias is found, then it is probably prudent to supplement the usual analyses with additional analyses that are designed to address selection bias, and still provide valid treatment comparisons. In Chapter 7 we mentioned using P{E} itself as a covariate, albeit a somewhat unusual one, for this purpose. It is useful to graph outcomes by P{E} (RPS) values separately for each treatment group, but overlaid on the same graph, as was done in Chapters 6 and 7 (the Berger–Exner graph). This provides a visual display that can detect selection bias and correct for it, as well as help to provide an estimate of G, the cutoff used for determining when to bias the patient selection. If, for example, the graph of outcome by P{E} is flat within each treatment group, then this suggests that there was no selection bias, and $G = 1.00$. If the graph is flat in the middle but rises to the right of some point (say 0.75) and drops to the left of the complementary point $(1 - 0.75 = 0.25)$, then this suggests that $G = 0.75$. Looking for a symmetric change-point, or an increase in the mean response for RPS values of both $0.5 - k$ and $0.5 + k$ for some value of k, with flat patters on each of the three regions $[0.0, 0.5 - k)$, $[0.5 - k, 0.5 + k]$, and $(0.5 + k, 1.0]$, is an alternative to the likelihood approach discussed in Section 7.3.2. Specifically, G would be estimated as $0.5 + k$. For example, if the graph is flat on $[0.0, 0.34)$, jumps to a higher flat pattern on $[0.34, 0.66]$, and jumps again to another flat pattern on $(0.66, 1.00]$, then this would suggest that there were biased allocations when P{E} was $1/3$ or under, or $2/3$ or over. We would then estimate G to be $2/3$. But if there are flat regions consisting of $\{0.0\}$, $(0.0, 1.00)$, and $\{1.00\}$, then we would infer instead that $G = 0.99$, and that the only biased allocations occurred when P{E} $= 0.0$ or 1.0. Unfortunately, the estimation of G by this technique is subjective, but this can still be a useful piece of information in summarizing any selection bias found.

We also discussed an adjustment technique based on the likelihood (Ivanova *et al.*, 2005). These methods do not require data on patients who were not randomized, but if such data are available, then other analyses can be conducted as well. For example, one could conduct an intent-to-randomize analysis, in which all patients screened, whether randomized or not, are included and assigned to the treatment group they would have been assigned to had they been randomized and had all previously screened patients also been randomized (Berger and

Exner, 1999). Of course, some screened patients are denied enrollment for legitimate reasons having nothing to do with selection bias, so if there is some way to distinguish the two causes for denial (legitimate and selection bias), and determine the range of P{E} values for which each patient would have been enrolled, then this can be incorporated into an appropriate permutation test that adjusts for any selection bias found without assuming that all screened patients would have been enrolled had there not been any selection bias.

If one does adjust for P{E}, the reverse propensity score (RPS), then this might have implications for which other covariates should be included in the model. It may turn out that the best covariates (in terms of prognostic ability for the primary efficacy endpoint) are the ones most associated with P{E}, so inclusion of P{E} in the model might obviate the need for including some other covariates. Clearly, more research is needed in determining how best to handle selection bias at the analysis stage, but what is clear is that these efforts should focus on detection and correction. Berger and Christophi (2003) listed items to be reported in randomized trials in their Table 8.4.

Many of these points are relevant to the analysis stage of a trial, and are included in Table 8.5.

If all of these measures are taken, or at least addressed, then the result will be a climate much more hostile to selection bias, because selection bias would be much more easily detected. It may be overly optimistic to hope that these steps would bring an end to selection bias as we know it, but it is reasonable to suppose that they might be the beginning of the end. That is, greater respect for selection bias, and other biases as well, may lead to further research into the management of these problems, and such further research may result in a complete elimination of the problem. Until that day comes, about all we can do is keep an open mind regarding these biases. This open mind can easily be misconstrued to mean a trusting mind towards those who would dismiss these biases, or who express indignation at having their data subjected to analyses aimed at detecting these biases. In fact the open mind refers to allowing for the possibility, and not being quick to dismiss the problem based on the argument that this has never been a problem in the past. This argument uses a form of circular logic (Rips, 2002), in that the lack of a problem means that no problem manifested itself. This would cause the lack of scrutiny that would in turn allow the problem to go on undetected. The methods discussed

Table 8.4   What to report in randomized clinical trials to control selection bias

Concern	Report
Different Allocation Discretion	Planned allocation proportions
	Number of screened and randomized patients by the group to which they were or would have been randomized had they been randomized
Deferred Enrollment	List patients who were screened twice or more, or that there were none
Allocation Concealment	Specific means of concealing the future allocations
Predicted Allocations	Specific restrictions on the randomization (including block sizes)
	Specific methods of concealing the past allocations (masking)
	Evidence of unmasking (including differential rates of observable adverse events, any emergencies requiring intentional unmasking, and rates of correct treatment group guesses at de-briefing)
Baseline Imbalances	Compare baseline covariates across treatment groups
Selection Bias	Graph key covariates against P{active}, as in Berger and Exner (1999)
	Graph response against P{active} within each treatment group, per Berger and Exner (1999).
	List stratification errors (if any), or that there were none

in this book will allow researchers to take the more scientific 'trust but verify' approach.

8.4   ACTION POINTS BY PARTY

To this point we have discussed actions to be taken by those who design, conduct, and analyze randomized trials. Yet there are other parties who can (and should) also play a role in managing selection bias in randomized trials. Before we discuss what these actions are, we first

Table 8.5   Items to consider during the analysis stage of a randomized trial

---

Conduct and report the Berger–Exner test of selection bias if appropriate (Table 8.3).

Report the specific restrictions on the randomization.

Graph the key endpoints against P{E} by treatment group.

Report the associations between each key covariate and P{E}.

If covariate information is available for non-randomized patients, then analyze it.

If the trial was planned as masked, then comment on the success of masking.

Comment on the success of allocation concealment.

If selection bias is found, then present between-group analyses within levels of P{E}.

Consider using the likelihood approach to adjusting for selection bias.

Consider the intent-to-randomize approach (if response data are available for all screened patients).

If selection bias is found, then present an estimate of G and the most likely selection covariate.

---

need to characterize the present system for dealing with selection bias, so that we have a basis for comparison. In a word, the current system can be characterized as trust. Patients trust that their physicians make appropriate decisions regarding their medical care. These physicians in turn trust both the regulatory authorities and the journals to keep unsafe medicines off the market and deliver unbiased information, respectively. The regulatory authorities and the journals, in turn, trust the sponsors and authors to give a clear idea of what is going on with the treatments they have evaluated in hopes of approval and/or publication. And these sponsors trust the investigators to conduct the studies in a manner that does not introduce bias.

If investigators do introduce bias, perhaps selection bias, and nobody on the path from the investigator to the patient stops to ask the right questions, then it is clear that the patient may suffer from suboptimal health care. Rather than asking where the blame lies, it is best to ask how this situation can be avoided. In other situations, trust in those on the front line to do the right thing has been shown to be a poor substitute for systems that ensure compliance. For example, in many jurisdictions it is no longer optional to wear a seat belt while operating a motor vehicle – rather, this is the law. In others, it is no longer legal to speak on a cell phone while operating a motor vehicle. Trusting drivers to make the right decision was not producing the desired result, and so stronger measures were required – and taken. One could also trust physicians to just be careful and not make mistakes, yet it was recognized (Wachter and Shojania, 2004) that part of the problem causing medical errors was poor procedures themselves (as opposed to adherence to these procedures), and so efforts are geared towards procedures that will prevent even the possibility of these errors. Likewise, one could trust sponsors to be forthcoming with all relevant information concerning the treatments they study, yet it was found that the suppression of negative findings could sometimes lead to distorted conclusions. Hence, rather than urging sponsors to follow their conscience, some medical journals are now stating as a condition of publication of a trial that it was pre-registered in a public trial registry (DeAngelis *et al.*, 2004).

These situations discussed above illustrate that the "trust, don't ask, and don't tell" approach takes one only so far in bringing about the desired result. And yet this is exactly what is currently being done to manage selection bias. It would seem that any change would be a change for the better. But perhaps the best changes would be those that bring about more attention to the problem. Specifically, then, we would call on investigators to resist the temptation to predict future allocations, and for sponsors to design trials that would resist such attempts. We call on journals to require authors to make publicly available the full data sets on which publications are based. We call on regulatory authorities to delve deeper into quality checks of individual trials, and specifically to check for selection bias among the myriad

Table 8.6 Actions To Be Taken by Various Involved Parties

Investigators	Do not try to predict future allocations.
Sponsors	Design trials that are less susceptible to selection bias.
Journals	Require full data to be made available for publications.
Regulatory Authorities	Add checks of selection bias to set of routine checks.
Meta-analysts	Consider selection bias when weighting studies.
Consumers	Demand discussions of biases before accepting results.
Methodologists	Develop better methods to manage selection bias.

other biases routinely checked. We call on meta-analysts and policy-makers to weight studies by their quality, and to include assessments of selection bias in these evaluations of quality of individual trials. We call on consumers of medical research (including health maintenance organizations) to be skeptical, and to demand information regarding biases that may have affected the quality of the studies. We call on funding agencies to do the same, but before-the-fact. Finally, we call on methodologists to develop better methods to prevent, detect, and adjust for selection bias in randomized trials. See Table 8.6 for a summary of these various action points.

References

Alexander, F. E., Anderson, T. J., Brown, H. K., Forrest, A. P. M., Hepburn, W., Kirkpatrick, A. E., Muir, B. B., Prescott, R. J., and Smith, A (1999). 14 years of follow-up from the Edinburgh Randomised Trial of Breast-Cancer Screening. *Lancet*, **353**, 1903–1908.

Altman, D. G. (1985). Comparability of randomized groups. *The Statistician*, **34**, 125–136.

Asarnow, J. R., Jaycox, L. H., Duan, N., LaBorde, A. P., Rea, M. M., Murray, P., Anderson, A., Landon, C., Tang, L., and Wells, K. B. (2005). Effectiveness of a quality improvement intervention for adolescent depression in primary care clinics. *JAMA*, **3**, 311–319.

Atkinson, A. C. (2001). Authors' reply. *Statistics in Medicine*, **20**, 816–818.

Ayala, E. and MacKillop, N. (2001). When to break the blind. *Applied Clinical Trials*, **10**(11), 61–62.

Bailar, J. C. (1976). Bailar's laws of data analysis. *Clinical Pharmacology and Therapy*, **20**, 113–120.

Bailar, J. C. and MacMahon, B. (1997). Randomization in the Canadian National Breast Screening Study: A review for evidence of subversion. *Canadian Medical Association Journal*, **156**, 193–199.

Bailenson, J. N. and Rips, L. J. (1996). Informal reasoning and burden of proof. *Applied Cognitive Psychology*, **10**, S3–S16.

Bakos, O. and Backstrom, T. (1987). Induction of labor: A prospective, randomized study into amniotomy and oxytocin as induction methods in a total unselected population. *Acta Obstetrica et Gynecologica Scandinavica*, **66**, 537–541.

Barraclough, K. (2003). There is no evidence that . . . . *British Medical Journal*, **326**, 1095.

Beller, E. M., Gebski, V., and Keech, A. C. (2002). Randomization in clinical trials. *Medical Journal of Australia*, **177**, 10, 565–567.

Berger, V. W. (1999). FDA Product Approval Information – Licensing Action: Statistical Review. http://www.fda.gov/cder/biologics/review/etanimm052799r2.pdf (accessed February 28, 2005).

Berger, V. W. (2000). Pros and cons of permutation tests in clinical trials. *Statistics in Medicine*, **19**, 1319–1328.

Berger, V. W. (2001). The *p*-value interval as an inferential tool. *The Statistician*, **50**, 79–85.

Berger, V. W. (2005a). Quantifying the magnitude of baseline covariate imbalance resulting from selection bias in randomized clinical trials. *Biometrical Journal*. In press.

Berger, V. W. (2005b). The reverse propensity score to correct for selection bias in randomized trials. *Statistics in Medicine*, **24**. In press.

Berger, V. W. and Bears, J. (2003). When can a clinical trial be called 'randomized'? *Vaccine*, **21**, 468–472.

Berger, V. W. and Christophi, C. A. (2003). Randomization technique, allocation concealment, masking, and susceptibility of trials to selection bias. *Journal of Modern Applied Statistical Methods*, **2**(1), 80–86.

Berger, V. W. and Exner, D. V. (1999). Detecting selection bias in randomized clinical trials. *Controlled Clinical Trials* **20**, 319–327.

Berger, V. W. and Weinstein, S. (2004). Ensuring the comparability of comparison groups: Is randomization enough? *Controlled Clinical Trials*, **25**, 515–524.

Berger, V. W., Permutt, T. and Ivanova, A. (1998). The convex hull test for ordered categorical data. *Biometrics*, **54**, 1541–1550.

Berger, V. W., Lunneborg, C., Ernst, M. D. and Levine, J. G. (2002). Parametric analyses in randomized clinical trials. *Journal of Modern Applied Statistical Methods*, **1**(1), 74–82.

Berger, V. W., Ivanova, A. and Deloria Knoll, M. (2003a). Minimizing predictability while retaining balance through the use of less restrictive randomization procedures. *Statistics in Medicine*, **22**, 3017–3028.

Berger, V. W., Rezvani, A. and Makarewicz, V (2003b). Direct effect on validity of response run-in selection in clinical trials. *Controlled Clinical Trials*, **24**, 156–166.

Berger, V. W., Zhou, Y. Y., Ivanova, A. and Tremmell, L. (2004). Adjusting for ordinal covariates by inducing a partial ordering. *Biometrical Journal*, **46**, 48–55.

Bezwoda, W. R., Seymour, L., and Dansey, R. D. (1995). High-dose chemotherapy with hematopoietic rescue as primary treatment for metastatic breast cancer: A randomized trial. *Journal of Clinical Oncology*, **13**, 2483–2489.

Bird, S. M. (2001). Dissemination of decisions on interim analyses needs wider debate. *British Medical Journal*, **323**, 1424.

Blackwell, D. and Hodges, J. L. (1957). Design for the control of selection bias. *Annals of Mathematical Statistics*, **28**, 449–460.

Bloom, J. M. (2002). Letter to the Editor. *Lancet*, **359**, 2201.

Boyd, N. F. (1997). The review of randomization in the Canadian National Breast Screening Study. *Canadian Medical Association Journal*, **156**, 207–209.

Brauer, C. (2004). Failure Analysis of Clinical Studies with Medical Devices. Presentation at the Henry Stewart Conference, Gaithersburg, MD, USA, October 27, 2004.

Breslow, N. E. and Day, N. E. (1980). *Statistical Methods in Cancer Research, Volume 1. The Analysis of Case–Control Studies*, IARC Scientific Publication no. 32. Lyon: International Agency for Research on Cancer.

Burgess, D. C., Gebski, V. J. and Keech, A. C. (2003). *Medical Journal of Australia*, **179**, 2, 105–107.

Carleton, R. A., Sanders, C. A. and Burack, W. R. (1960). Heparin administration after acute myocardial infarction. *New England Journal of Medicine*, **263**, 1002–1005.

Carroll, K. M., Rounsaville, B. J. and Nich, C. (1994). Blind man's bluff: Effectiveness and significance of psychotherapy and pharmacotherapy blinding procedures in a clinical trial. *Journal of Consulting and Clinical Psychology*, **62**, 276–280.

CASS Investigators (1984). Coronary Artery Surgery Study (CASS): A randomized trial of coronary bypass surgery. Comparability of entry characteristics and survival in randomized patients and nonrandomized patients meeting randomized criteria. *Journal of the American College of Cardiology*, **3**, 114–128.

Chalmers, T. C. (1990). Discussion of 'Biostatistical collaboration in medical research' by Jonas H. Ellenberg. *Biometrics*, **46**, 20–22.

Chalmers, T. C., Celano, P., Sacks, H. S. and Smith, H. (1983). Bias in treatment assignment in controlled clinical trials. *New England Journal of Medicine*, **309**, 1358–1361.

Chen, Y. P. (1999). Biased coin design with imbalance intolerance. *Communications in Statistics Stochastic Models*, **15**, 953–975.

Clarke, M. (2002). Last moment randomization and concealment. *British Medical Journal*, **323**. http://bmj.com/cgi/eletters/323/7310/446 (accessed April 10, 2002).

Cohen, M. M., Kaufert, P. A., MacWilliam, L. and Tate, R. B. (1996). Using an alternative data source to examine randomization in the Canadian National Breast Screening Study. *Journal of Clinical Epidemiology*, **49**, 1039–1044.

COMET Study Group (2001). Effect of low-dose mobile versus traditional epidural techniques on mode delivery: A randomized controlled trial. *Lancet*, **358**, 19–23.

Day, S. (1998). Blinding or Masking. In P. Armitage and T. Colton (eds), *The Encyclopedia of Biostatistics*, Volume 1, pp. 410–417. Chichester: John Wiley & Sons, Ltd.

DeAngelis, C. D., Drazen, J. M., Frizelle, F. A., *et al.* (2004). Clinical trial registration: A statement from the International Committee of Medical Journal Editors. *JAMA* **292**, 1363–1364.

Dupin-Spriet, T., Fermanian, J. and Spriet, A. (2004). Quantification of predictability in clinical trials using block randomization. *Drug Information Journal*, **38**, 127–133.

Ellenberg, S. S., Epstein, J. S., Fratantoni, J. C., *et al.* (1994). A trial of RSV immune globulin in infants and young children: The FDA view. *New England Journal of Medicine*, **331**, 203–205.

ENRICHD Investigators (2003). Effects of treating depression and low perceived social support on clinical events after myocardial infarction. *JAMA*, **289**, 23, 3106–3116.

Fayers, P. M. and Sprangers, M. A. (2002). Understanding self-rated health. *Lancet*, **359**, 187–188.

Feigenbaum, S. and Levy, D. M. (1996). The technological obsolescence of scientific fraud. *Rationality and Society*, **8**, 261–276.

Fentiman, I. S., Rubens, R. D. and Hayward, J. L. (1983). Control of pleural effusions in patients with breast cancer. *Cancer*, **52**, 737–739.

Fergusson D., Aaron S. D., Guyatt G. and Hebert P. (2002). Post-randomisation exclusions: The intention to treat principle and excluding patients from analysis. *British Medical Journal*, **325**, 652–654.

Fergusson, D., Glass, K. C., Waring, D. and Shapiro, S. (2004a). Turning a blind eye: The success of blinding reported in a random sample of randomised, placebo-controlled trials. *British Medical Journal*, **328**, 432–440.

Fergusson, D., Glass, K. C., Waring, D., and Shapiro, S. (2004b). Turning a blind eye: Authors' reply. *British Medical Journal*, **328**, 1136.

Fiellin, D. A., O'Connor, P. G., Chawarski, M., *et al.* (2001). Methadone maintenance in primary care – a randomized controlled trial. *Journal of the American Medical Association*, **286**, 1724–1731.

Follmann, D. and Proschan, M. (1994). The effect of estimation and biasing strategies on selection bias in clinical trials with permuted blocks. *Journal of Statistical Planning and Inference* **39**, 1–17.

Frane, J. W. (1998). A method of biased coin randomization, its implementation, and its validation. *Drug Information Journal*, **32**, 423–432.

Frangakis C. E. and Rubin D. B. (2002). Principal stratification in causal inference. *Biometrics*, **58**, 21–29.

Gallin, J. I., Alling, D. W., Malech, H. L., Wesley, R., *et al.* (2003). Itraconazole to prevent fungal infections in chronic granulomatous disease. *New England Journal of Medicine*, **348**, 2416–2422.

Gansky, S. A. and Koch, G. G. (2001). Assessing bias of multicenter trials with incomplete treatment allocation. *Journal of Statistical Planning and Inference*, **96**, 83–107.

Geller, N. L., Sorlie, P., Coady, S., Fleg, J. and Friedman, L. (2004). Limited access data sets from studies funded by the National Heart, Lung, and Blood Institute. *Clinical Trials*, **1**, 517–524.

Gerdesmeyer, L., Wagenpfeil, S., Haake, M., Maier, M., Loew, M., Wortler, K. Lampe, R., Seil, R., Handle, G., Gassel, S. and Rompe, J. D. Extracorporeal shock wave therapy for the treatment of chronic calcifying tendonitis of the rotator cuff. *JAMA*, **290**, 19, 2573–2580.

Gleason, D. F. (1977), Histologic grading and clinical staging of prostatic carcinoma. In M. Tannenbaum (ed.), *Urologic Pathology: The Prostate*, pp. 171–197. Philadelphia: Lea & Febiger.

Gotzsche P. C. and Olsen O. (2000a). Is screening for breast cancer with mammography justifiable? *Lancet*, **355**, 129–134.

Gotzsche P. C. and Olsen O. (2000b). Screening mammography re-evaluated: Reply. *Lancet*, **355**, 752.

Green, S. B. and Byar, D. P. (1984). Using observational data from registries to compare treatments: The fallacy of omnimetrics. *Statistics in Medicine*, **3**, 361–370.

Greenhouse, S. W. (2003). The growth and future of biostatistics. *Statistics in Medicine*, **22**, 3323–3335.

Greenland, S. (1998). Induction versus Popper: Substance versus semantics. *International Journal of Epidemiology*, **27**, 543–548.

Greenland, S. and Robins, J. M. (1986). Identifiability, exchangeability, and epidemiological confounding. *International Journal of Epidemiology* **15**, 413–419.

Groothuis, J. R., Simoes, E. A. F., Levin, J. M., *et al.* (1993). Prophylactic administration of respiratory syncytial virus immune globulin to high-risk infants and young children. *New England Journal of Medicine*, **329**, 1524–1530.

Guyatt, G., Cook, D., Devereaux, P. J., Meade, M. and Straus, S. (2002). Therapy. Section IBI (pages 55–79) of The Users' Guide to the Medical Literature: A Manual for Evidence-Based Clinical Practice, G. Guyatt and D. Rennie, Editors, AMA Press, Chicago.

Hallstrom, A. and Davis, K. (1988). Imbalance in treatment assignments in stratified blocked randomization. *Controlled Clinical Trials*, **9**, 375–382.

Hansen, J. B., Smithers, B. M., Schache, D., Wall, D. R., Miller, B. J. and Menzies, B. L. (1996). Laparoscopic versus open appendectomy: Prospective randomized trial. *World Journal of Surgery*, **20**, 17–21.

Heckman, J. J., Ichimura, H., Smith, J. and Todd, P. (1996). Sources of selection bias in evaluating social programs: An interpretation of conventional measures and evidence on the effectiveness of matching as a program evaluation method. *Proceedings of the National Academy of Sciences*, **93**, 13 416–13 420.

Hollenbeak, C. S., Murphy, D., Dunagan, W. C. and Fraser V. J. (2002). Nonrandom selection and the attributable cost of surgical-site infections. *Infection Control and Hospital Epidemiology*, **23**, 177–182.

HOPE Investigators (2000). Effect of an Angiotensin-Converting-Enzyme Inhibitor, Ramipril, on cardiovascular events in high-risk patients. *New England Journal of Medicine*, **342**, 3, 145–153.

Hughes, J. P. and Richardson, B. A. (2000). Analysis of a randomized trial to prevent vertical transmission of HIV-1. *Journal of the American Statistical Association*, **95**, 1032–1043.

Humphrey, L. L., Helfand, M., Chan, B. K. S. and Woolf, S. H. (2002). Breast cancer screening: A summary of the evidence for the US Preventive Services Task Force. *Annals of Internal Medicine*, **137**(5, Part 1), 347–360.

Hunkeler, E. M., Meresman, J. F., Hargreaves, W. A., *et al.* (2000). Efficacy of nurse telehealth care and peer support in augmenting treatment of depression in primary care. *Archives of Family Medicine*, **9**, 700–708.

Ivanova, A., Barrier, R. C., and Berger, V. W. (2005). Adjusting for observable selection bias in block randomized trials. *Statistics in Medicine*, **24**. In press.

Jones, J. W., McCullough, L. B. and Richman, B. W. (2003). The ethics of sham surgery in research. *Journal of Vascular Surgery*, **37**, 482–483.

Jordhoy, M. S., Fayers, P. M., Ahlner-Elmqvist, M. and Kaasa, S. (2002). Lack of concealment may lead to selection bias in cluster randomized trials of palliative care. *Palliative Medicine*, **16**, 43–49.

Keirse, M. J. N. C. (1988). Amniotomy or oxytocin for induction of labor. *Acta Obstetrica et Gynecologica Scandinavica*, **67**, 731–735.

Kennedy, A. and Grant, A. (1997). Subversion of allocation in a randomized controlled trial. *Controlled Clinical Trials*, **18**(3S), 77S–78S.

Kjaergard, L. L. and Als-Nielsen, B. (2002). Association between competing interests and authors' conclusions: Epidemiological study of randomised clinical trials published in the *BMJ. British Medical Journal*, **325**, 249–252.

Kolata, G. (2002). Different conclusion from the same study. *New York Times*, April 9, Science Section.

Kroenke, K., West, S. L., Swindle, R., *et al.* (2001). Similar effectiveness of paroxetine, fluoxetine, and sertraline in primary care – a randomized trial. *Journal of the American Medical Association*, **286**, 2947–2955.

Kunz, R. and Oxman, A. D. (1998). The unpredictability paradox: Review of empirical comparisons of randomised and nonrandomised clinical trials. *British Medical Journal*, **317**, 1185–1190.

Kurland, J. (2002). The heart of the precautionary principle in democracy. *Public Health Reports*, **117**, 498–500.

Kuznetsova, O. M. (2002). Why permutation is even more important in IVRS drug codes schedule generation than in patient randomization schedule generation. *Controlled Clinical Trials*, **22**, 69–71.

Leber, P. D. and Davis, C. S. (1998). Threats to the validity of clinical trials employing enrichment strategies for sample selection. *Controlled Clinical Trials*, **19**, 178–187.

Lippman, S. M., Lee, J. J., Kurp, D. D., *et al.* (2001). Randomized Phase III Intergroup Trial of Isotretinoin to Prevent Second Primary Tumors in Stage I Non-Small-Cell Lung Cancer. *Journal of the National Cancer Institute*, **93**, 605–618.

Lindholm, L. H., Ibsen, H., Dahlof, B., *et al.* (2002). Cardiovascular morbidity and mortality in patients with diabetes in the Losartan Intervention for Endpoint Reduction in Hypertension Study (LIFE): A randomized trial against atenolol. *Lancet*, **359**, 1004–1010.

Lovell, D. J., Giannini, E. H., Reiff, A., *et al.* (2000). Etanercept in children with polyarticular juvenile rheumatoid arthritis. *New England Journal of Medicine*, **342**, 763–769.

Mack, T. M., Pike, M. C., Henderson, B. E., Pfeffer, R. I., Gerkins, V. R., Arthur, M., and Brown, S. E. (1976). Estrogens and endometrial cancer in a retirement community. *New England Journal of Medicine*, **294**, 1262–1267.

Madersbacher, S., Thalman, G. N., Fritsch, J. C. and Studer, U. E. (2004). Is eligibility for a chemotherapy protocol a good prognostic factor for invasive bladder cancer after radical cystectomy? *Journal of Clinical Oncology*, **22**, 20, 4103–4108.

Mann, J. (2004). A critique of the reanalysis of the NINDS trial. http://www.homestead.com/emguidemaps/files/IngallCritique.htm, accessed November 15, 2004.

Marcus, S. M. (2001). A sensitivity analysis for subverting randomization in controlled trials. *Statistics in Medicine* **20**, 545–555.

Mark, D. H. (1997). Interpreting the term selection bias in medical research. *Family Medicine*, **29**(2), 132–136.

Martin, J. L. R., Barbanoj, M. J., Schlaepfer, T. E., Thompson, E., Perez, V. and Kulisevsky, J. (2003). Repetitive transcranial magnetic stimulation for the treatment of depression: systematic review and meta-analysis. *The British Journal of Psychiatry*, **182**, 480–491.

Matts, J. P. and Lachin, J. M. (1988). Properties of permuted block randomibreakzation in clinical trials. *Controlled Clinical Trials*, **9**, 327–344.

Matts, J. P. and McHugh, R. B. (1983). Conditional Markov chain design for accrual clinical trials. *Biometrical Journal*, **25**, 563–577.

McCulloch, P., Taylor, I., Sasako, M. Lovett, B. and Griffin, D. (2002). Randomized trials in surgery: problems and possible solutions. *British Medical Journal*, **324**, 1448–1451.

McEntegart, D. (2003). Forced randomization when using interactive voice response systems. *Applied Clinical Trials* **12**(10), 50–58.

McKay, B., Bar-Natan, D., Bar-Hillel, M. and Kalai, G. (1999). Solving the Bible code puzzle. *Statistical Science*, **14**(2), 150–173.

Miettinen, O. S. and Cook, E. F. (1981). Confounding: Essence and detection. *American Journal of Epidemiology*, **114**, 593–603.

Mitchell, J. R. A. (1981). Timolol after myocardial Infarction: An answer or a new set of questions? *British Medical Journal*, **282**, 1565–1570.

Moher, D., Pham, B., Jones, A., *et al.* (1998). Does quality of reports of randomised trials affect estimates of intervention efficacy reported in meta-analysis? *Lancet*, **352**, 609–613.

Morton, V. and Torgerson, D. J. (2003). Effect of regression to the mean on decision making in health care. *British Medical Journal*, **326**, 1083–1087.

Moses L. E. (1995). Measuring effects without randomized trials? Options, problems, challenges. *Medical Care*, **33**(4), AS8–AS14.

Mountain, C. F. and Gail, M. H. (1981). Surgical adjuvant intrapleural BCG treatment for stage I non-small cell lung cancer: Preliminary report of the National Cancer Institute Lung Cancer Study Group. *Journal of Thoracic and Cardiovascular Surgery*, **82**, 649–657.

Nance, D. A. (1998). Evidential completeness and the burden of proof. *Hastings Law Journal*, **49**, 621–650.

NINDS rt-PA Stroke Study Group (1995). Tissue plasminogen activator for acute ischemic stroke. *New England Journal of Medicine*, **24**, 333, 1581–1587.

Olschewski M., Schumacher M., and Davis K. B. (1992). Analysis of randomized and nonrandomized patients in clinical trials using the comprehensive cohort follow-up study design. *Controlled Clinical Trials*, **13**, 226–239.

Ornish, D., Scherwitz, L. W., Billings, J., *et al.* (1998). Intensive lifestyle changes for reversal of coronary heart disease. *Journal of the American Medical Association*, **280**, 2001–2007.

Oxtoby, A., Jones, A., and Robinson, M. (1989). Is your 'double-blind' design truly double-blind? *British Journal of Psychiatry*, **155**, 700–701.

Paradise, J. L., Bluestone, C. D., Bachman, R. Z., *et al.* (1984). Efficacy of tonsillectomy for recurrent throat infection in severely affected children. *New England Journal of Medicine*, **310**, 674–683.

Penston, J. (2003). *Fiction and Fantasy in Medical Research: The Large-Scale Randomised Trial*. The London Press, London.

Peto, R. (1999). Failure of randomisation by 'sealed' envelope. *Lancet*, **354**, 73.

Peto, R., Collins, R. and Gray, R. (1995). Large-scale randomized evidence: Large, simple trials and overviews of trials. *Journal of Clinical Epidemiology*, **48**, 23–40.

Pocock, S. J. and Lagakos, S. W. (1982). Practical experience of randomization in cancer trials: An international survey. *British Journal of Cancer* **46**, 368–375.

Pocock, S. J. and Simon, R. (1975). Sequential treatment assignment with balancing for prognostic factors in the controlled clinical trial. *Biometrics*, **31**, 103–115.

POTS Team (2004). Cognitive-behavior therapy, sertraline, and their combination for children and adolescents with obsessive-compulsive disorder: the Pediatric OCD Treatment Study (POTS) randomized controlled trial. *JAMA*, **292**, 16, 1969–1976.

Prorok, P. C., Hankes, B. F. and Bundy, B. N. (1981). Concepts and problems in the evaluation of screening programs. *Journal of Chronic Diseases*, **34**, 159–171.

Proschan, M (1994). Influence of selection bias on Type I error rate under random permuted block designs. *Statistica Sinica*, **4**, 219–231.

Psaty, B. M. and Rennie, D. (2003). Stopping medical research to save money: A broken pact with researchers and patients. *Journal of the American Medical Association*, **289**, 2128–2131.

Psaty, B., Furberg, C., Pahor, M., *et al.* (2000). National guidelines, clinical trials, and quality of evidence. *Archives of Internal Medicine*, **160**, 2577–2580.

Reynolds, S. M. (2004). ORI findings of scientific misconduct in clinical trials and publicly funded research, 1992–2002. *Clinical Trials*, **1**, 509–516.

Rips, L. J. (2002). Circular reasoning. *Cognitive Science*, **26**, 767–795.

Rosenbaum, P. R. (2001). Replicating effects and biases. *American Statistician*, **55**, 223–227.

Rosenberger, W. and Lachin, J. M. (2002) *Randomization in Clinical Trials: Theory and Practice*, New York: John Wiley and Sons, Inc.

Rosner, D. and Markowitz, G. (2002). Industry challenges to the principle of prevention in public health: The precautionary principle in historic perspective. *Public Health Reports*, **117**, 501–512.

Rothman, K. J. (1977). Epidemiologic methods in clinical trials. *Cancer*, **39**, 1771–1775.

Rubin, D. B. (1977). Assignment to treatment group on the basis of a covariate. *Journal of Educational Statistics*, **2**(1), 1–26.

Runde, J (1996). On Popper, probabilities, and propensities. *Review of Social Economy*, **54**, 465–485.

Schor, S. (1971). The University Group Diabetes Program: A statistician looks at the mortality results. *Journal of the American Medical Association*, **217**, 1671–1675.

Schulz, K. F. (1995a). Subverting randomization in controlled trials. *Journal of American Medical Association* **274**, 1456–1458.

Schulz, K. F. (1995b). Unbiased research and the human spirit: The challenges of randomized controlled trials. *Canadian Medical Association Journal*, 153, 783–786.

Schulz, K. F. (1996). Randomised trials, human nature, and reporting guidelines. *Lancet*, **348**, 596–598.

Schulz, K. F. and Grimes, D. A. (2002a). Allocation concealment in randomised trials: Defending against deciphering. *Lancet*, **359**, 614–618.

Schulz, K. F. and Grimes, D. A. (2002b). Generation of allocation sequences in randomized trials: Chance, not choice. *Lancet*, **359**, 515–519.

Schulz, K. F., Chalmers, I., Hayes, R. J. and Altman, D. G. (1995). Empirical evidence of bias: Dimensions of methodological quality associated with estimates of treatment effects in controlled trials. *Journal of the American Medical Association*, **273**, 408–412.

Senn, S. J. (1991). Falsificationism and clinical trials. *Statistics in Medicine*, **10**, 1679–1692.

Senn, S. J. (1994). Testing for baseline balance in clinical trials. *Statistics in Medicine*, **13**, 1715–1726.

Senn, S. (1995). A personal view of some controversies in allocating treatment to patients in clinical trials. *Statistics in Medicine*, **14**, 2661–2674.

Senn, S. (1997), *Statistical Issues in Drug Development*. Chichester: John Wiley & Sons, Ltd.

Senn, S. (2000). Consensus and controversy in pharmaceutical statistics. *The Statistician*, **49**, 135–176.

Singh, H. P. and Mathur, N. (2004). Unknown repeated trials in the unrelated question randomized response model. *Biometrical Journal* **46**, 375–378.

Sleight, P., Pogue, J., and Yusuf, S. (2002). Author's reply. *Lancet*, **359**, 2118.

Sleight, P., Yusuf, S., Pogue, J., Tsuyuki, R., Diaz, R., and Probstfield, J. (2001). Blood-pressure reduction and cardiovascular risk in HOPE study. *Lancet*, **358**, 2130–2131.

Smith, R. (2003). Do Patients need to read research? *British Medical Journal*, 326, 1307.

Soares, J. F. and Wu, C. F. J. (1983). Some restricted randomization rules in sequential designs. *Communications in Statistics Theory and Methods*, **12**, 2017–2034.

Socinski, M. A., Ivanova, A., *et al.* (submitted). A randomized phase II trial comparing every-3-week carboplatin/paclitaxel with every-3-week carboplatin and weekly paclitaxel in advanced non-small cell lung cancer. *JCO*, submitted.

Song, J., Belin, T. R., Lee, M. B., Gao, X. and Rotheram-Borus, M. J. (2001). Handling baseline differences and missing items in a longitudinal study of HIV risk among runaway youths. *Health Services & Outcomes Research Methodology*, **2**, 317–329.

Stevenson, H. C. and Davis, G. (1994). Impact of culturally sensitive AIDS video education on the AIDS risk knowledge of African American adolescents. *AIDS Education and Prevention*, **6**, 40–52.

Stigler, S. M. (1969). The use of random allocation for the control of selection bias. *Biometrika*, **56**, 3, 553–560.

Stoto, M. A. (2002). The precautionary principle and emerging biological risks: Lessons from swine flu and HIV in blood products. *Public Health Reports*, **117**, 546–552.

Swingler, G. H. and Zwarenstein, M. (2000). An effectiveness trial of a diagnostic test in a busy outpatients department in a developing country: Issues around allocation concealment and envelope randomization. *Journal of Clinical Epidemiology*, **53**, 702–706.

Tarone R. E. (1995). The excess of patients with advanced breast cancer in young women screened with mammography in the Canadian National Breast Screening Study. *Cancer*, **75**, 997–1003.

Taylor, R. (2002). Blood pressure and cardiovascular risk in the HOPE study. *Lancet*, **359**, 2117.

ter Riet, G. and Kessels, A. G. H. (1995). Restricted randomization in randomized controlled trials. *Journal of the American Medical Association*, **274**, 1835.

Torgerson, D. J. and Roberts, C. (1999). Randomisation methods: Concealment. *British Medical Journal*, **319**, 375–376.

Vamvakas E. C. (2000). Evaluation of clinical studies of the efficacy of therapeutic apheresis. *Journal of Clinical Apheresis*, **15**, 6–17.

Van den Brink, W., Hendriks, V. M., Blanken, P., Koeter, M. W. J., van Zwieten, B. J. and van Ree, J. M. (2003). Medical prescription of heroin to treatment resistant heroin addicts: Two randomised controlled trials. *British Medical Journal*, **327**, 310–315.

Van Dijk, D., Jansen, E. W. L., Hijman, R, *et al.* (2002). Cognitive outcome after off-pump and on-pump coronary artery bypass graft surgery – a randomized trial. *Journal of the American Medical Association*, **287**, 1405–1412.

Villar, J. and Carroli, G. (1996). Methodological issues of randomized clinical trials for the evaluation of reproductive health interventions. *Preventive Medicine* **25**, 365–375.

Wachter, R. M. and Shojania, K. (2004). *Internal Bleeding: The Truth Behind America's Terrifying Epidemic of Medical Mistakes*. Rugged Land, New York.

Walton, M. (1996). Clinical review for PLA 96-0350. Available at http://www.fda.gov/cder/biologics/products/altegen061896.htm, accessed November 15, 2004.

Wei, L. J. and Lachin, J. M. (1988). Properties of the urn randomization in clinical trials. *Controlled Clinical Trials*, **9**, 345–364.

Wei, L. and Zhang, J. (2001). Analysis of data with imbalance in the baseline outcome variable for randomized clinical trials. *Drug Information Journal*, **35**, 1201–1214.

Weiss, R. B., Gill, G. G. and Hudis, C. A. (2001). An on-site audit of the South African trial of high-dose chemotherapy for metastatic breast cancer and associated publications. *Journal of Clinical Oncology*, **19**, 2771–2777.

Author Index

Alexander, F. E. 52, 187
Als-Nielsen, B. 108, 192
Altman, D. G. 44, 126, 128, 187, 195
Asarnow, J. R. 81, 187
Atkinson, A. C. 112, 187
Ayala, E. 108, 175, 187

Backstrom, T. 45, 187
Bailar, J. C. 48, 49, 150, 153, 187
Bailenson, J. N. 41, 187
Bakos, O. 45, 187
Barraclough, K. 187
Bears, J. 4, 8, 10, 20, 45, 188
Beller, E. M. 9, 109, 110
Berger, V. W. 4, 8, 10, 11, 12, 13, 18,
    19, 20, 23, 24, 26, 30, 39, 42, 45,
    51, 76, 80, 81, 83, 85, 86, 94, 96,
    97, 98, 101, 102, 106, 107, 109,
    111, 112, 113, 116, 117, 119, 120,
    121, 122, 126, 131, 132, 133, 134,
    135, 136, 137, 138, 139, 140, 141,
    142, 144, 145, 146, 147, 149, 153,
    154, 158, 160, 162, 163, 164, 165,
    167, 168, 169, 175, 176, 178, 179,
    180, 181, 182, 183, 187, 188, 191
Bezwoda, W. R. 64, 188
Bird, S. M. 9, 19, 188
Blackwell, D. 43, 105, 115, 188
Bloom, J. M. 78
Boyd, N. F. 48, 49, 127, 189
Breslow, N. E. 6, 189

Brauer, C. 20
Bundy, B. N. 194
Burack, W. R. 189
Burgess, D. C. 125, 148
Byar, D. B. 18, 19, 94, 129, 191

Carleton, R. A. 43, 189
Carroli, G. 7, 46, 196
Carroll, K. M. 26, 189
CASS Investigators 50, 51, 189
Chalmers, I. 195
Chalmers, T. C. 21, 42, 189
Chen, Y. P. 117, 189
Christophi, C. A. 10, 11, 12, 13, 23, 24,
    26, 30, 81, 181, 188
Clarke, M. 11, 24, 189
Cohen, M. M. 48, 189
COMET Study Group 20, 189
Cook, E. F. 44, 193

Davis, C. S. 51, 192
Davis, G. 65, 195
Davis, K. B. 46, 47, 108, 193
Day, N. E. 6, 189
Day, S. 107, 189
DeAngelis, C. D. 184, 189
Deloria Knoll, M. 109, 116, 117, 188
Dupin-Spriet, T. 85, 87, 88, 89, 90, 190

Ellenberg, S. S. 47, 190
ENRICHD Investigators 80

Exner, D. V. 18, 19, 30, 39, 51, 80, 81, 107, 112, 122, 131, 132, 133, 134, 135, 136, 137, 138, 139, 140, 141, 142, 145, 146, 149, 154, 158, 160, 162, 163, 164, 167, 175, 178, 179, 180, 181, 182, 183, 188

Fayers, P. M. 7, 130, 190
Feigenbaum, S. 38, 41, 95, 190
Fentiman, I. S. 44, 45, 190
Fergusson, D. 25, 160, 176, 190
Fermanian, J. 190
Follmann, D. 113, 190
Frane, J. W. 110, 190
Frangakis C.E. 129, 190

Gail, M. H. 134, 135, 137, 193
Gallin, J. I. 8, 190
Gansky, S. A. 108, 190
Gebski, V. 9, 109, 110, 125, 148, 190
Geller, N. L. 83, 190
Gerdesmeyer, L. 80, 190
Gill, G. G. 196
Gleason, D. F. 128, 190
Gotzsche, P.C. 53, 159, 190
Grant, A. 49, 50, 192
Green, S. B. 18, 19, 94, 129, 191
Greenhouse, S. W. 42, 191
Greenland, S. 4, 129, 151, 152, 191
Grimes, D. A. 21, 23, 29, 30, 108, 122, 194, 195
Groothuis, J. R. 47, 191
Guyatt, G. 77, 190

Hallstrom, A. 46, 47, 108, 191
Hankes, B. F. 194
Hansen, J. B. 77, 190
Hayes, R. J. 195
Hayward, J. L. 190
Heckman J.J. 130, 191
Hodges, J. L. 43, 105, 115, 188
Hollenbeak C.S. 147, 191
HOPE Investigators 79, 191
Hudis, C. A. 196
Hughes, J. P. 63, 191

Humphrey, L. L. 53, 63, 191
Hunkeler, E. M. 29, 191

Ivanova, A. 42, 44, 102, 109, 116, 117, 118, 120, 145, 146, 147, 162, 165, 167, 168, 169, 180, 188, 191

Jones, A. 193
Jones, J. W. 13, 191
Jordhoy, M. S. 28, 52, 67, 192

Kalai, G. 193
Keech, A. C. 9, 109, 110, 125, 148
Keirse, M. J. N. C. 46, 192
Kennedy, A 49, 50, 192
Kessels, A. G. H. 30, 196
Kjaegard, L. L. 108, 192
Koch, G. G. 108, 190
Kolata, G. 53, 192
Kroenke, K. 80, 192
Kunz, R. 42, 192
Kurland, J. 149, 155, 192
Kuznetsova, O. M. 28, 192

Lachin, J. M. 22, 88, 113, 115, 117, 193, 194, 196
Lagakos, S. W. 110, 194
Leber, P. D. 51, 192
Levy, D. M. 38, 41, 95, 190
Lindholm, L. H. 78, 79, 192
Lippman, S. M. 29, 192
Lovell, D. J. 51, 52, 192

Mack, T. M. 6, 192
MacKillop, N. 108, 175, 187
MacMahon, B. 48, 150, 187
Madersbacher, S. 7, 10, 192
Makarewicz, V. 188
Mann, J. 74, 192
Marcus, S. M. 65, 192
Mark, D. H. 14, 18, 192
Markowitz, G. 149, 155, 194
Martin, J. L. R. 26, 193
Mathur, N. 176, 195
Matts, J. P. 88, 108, 193
McKay, B. 38, 193

McCulloch, P. 28, 193
McCullough, L. B. 191
McEntegart, D. 20, 108, 193
McHugh, R. B. 108, 193
Miettinen, O. S. 44, 193
Mitchell, J. R. A. 26, 75, 76, 158, 159, 193
Moher, D. 42, 193
Morton, V. 127, 193
Moses, L. E. 11, 129, 158, 193
Mountain, C. F. 134, 135, 137, 193

Nance, D. A. 39, 40, 193
NINDS rt-PA Stroke Study Group 74, 193

Olschewski M. 142, 160, 176, 193
Olsen, O. 53, 159, 190, 191
Ornish, D. 50, 193
Oxman, A. D. 42, 192
Oxtoby, A. 12, 25, 193

Paradise, J. L. 45, 193
Penston, J. 21, 26, 82, 151, 172, 173, 193
Permutt, T. 188
Peto, R. 53, 159, 172, 193
Pocock, S. J. 23, 24, 28, 110, 194
Pogue, J. 158, 195
POTS Team 80, 194
Prorok, P. C. 18, 194
Proschan, M. 51, 101, 113, 152, 190, 194
Psaty, B. M. 53, 63, 155, 194

Rennie, D. 155, 194
Reynolds, S. M. 43, 194
Rezvani, A. 188
Richardson, B. A. 63, 191
Richman, B. W. 192
Rips, L. J. 41, 181, 187, 194
Roberts, C. 17, 196
Robins, J. M. 4, 129, 152, 191
Rosenbaum P. R. 172, 194
Rosenberger, W. 88, 102, 113, 115, 117, 194

Rosner, D. 149, 155, 194
Rothman, K. J. 44, 123, 125, 158, 194
Rubens, R. D. 190
Rubin, D. B. 19, 129, 190, 194
Runde, J. 4, 152, 194

Sanders, C. A. 189
Schor, S. 43, 194
Schulz, K. F. 10, 12, 13, 17, 21, 23, 27, 28, 29, 30, 41, 42, 47, 108, 122, 194, 195
Schumacher M. 193
Senn, S. J. 17, 107, 112, 124, 125, 126, 132, 148, 153, 195
Shojania, K. 184, 196
Simon, R. 23, 25, 28, 194
Singh, H. P. 176, 195
Sleight, P. 79, 158, 195
Smith, A. 187
Smith, H. 189
Smith, J. 191
Smith, R. 3, 195
Soares, J. F. 117, 195
Socinski, M. A. 121, 195
Song, J. 65, 66, 67, 195
Sprangers, M. A. 7, 130, 190
Spriet, A. 190
Stevenson, H. C. 65, 195
Stigler, S. M. 115, 195
Stoto, M. A. 151, 195
Swingler, G. H. 63, 145, 160, 195

Tarone R.E. 127, 159, 196
Taylor, R. 79, 158, 196
ter Riet, G. 30, 196
Torgerson, D. J. 17, 127, 193, 196

Vamvakas E.C. 123, 196
Van den Brink, W. 72, 196
Van Dijk, D. 80, 196
Villar, J. 7, 46, 196

Wachter, R. M. 184, 196
Walton, M. 75, 196
Wei, L. 158, 196

Wei, L. J. 22, 196
Weinstein, S. 39, 42, 80, 83, 188
Weiss, R. B. 64, 196
Wu, C. F. J. 117, 195

Yusuf, S. 158, 195

Zhang, J. 158, 196
Zwarenstein, M. 63, 145, 160, 195

Subject Index

Advance Randomization 11, 14
Allocation Concealment 10, 12, 13, 14,
  18, 19, 21, 25, 27, 28, 29, 30, 31,
  33, 34, 35, 41, 42, 48, 53, 57, 63,
  64, 67, 72, 81, 107, 113, 123, 157,
  159, 161, 171, 172, 174, 175, 182,
  183
Allocation Discretion 19, 22, 172,
  182
Alternating Design 8, 20, 45
Attribution 4, 6, 18, 40, 124, 125,
  133

Baseline Covariate Imbalances 17, 18,
  24, 35, 41, 42, 43, 44, 45, 48, 50,
  51, 52, 53, 57, 63, 65, 66, 67, 69,
  70, 71, 75, 76, 78, 79, 85, 94, 95,
  96, 98, 100, 123, 124, 126, 127,
  128, 129, 130, 131, 134, 144, 148,
  149, 157, 158, 159, 160, 162, 163,
  173, 178, 182
Berger-Exner Test of Selection Bias 39,
  80, 132, 134, 146, 149, 154, 167,
  178, 179, 183
Berger-Exner Graph 132, 134, 135,
  136, 137, 138, 139, 140, 141, 142,
  162, 163, 164, 180
Bernoulli Trial 4
Biased Coin Design 117
Big Stick Rule 117
Burden of Proof 37, 38, 40, 41

Canadian National Breast Cancer
  Screening Study 48, 49, 54, 56, 58,
  60, 62, 127, 157, 159, 161
Captopril Prevention Project 53, 159
Case-Control Studies 6, 7
Causal Inference 4
Centralized Telephone System 49, 50
Chronological Bias 108, 109, 111, 112,
  115, 116, 117, 121, 122, 171, 172
Cluster Randomization 28, 52, 55, 65,
  67, 68
Comprehensive Cohort Follow-up Study
  142, 160, 176
Condition B 112, 113, 114, 115, 116,
  119, 121, 122
Condition F 113, 116
Condition T 112, 113, 116, 121, 122
Condition V 114, 116
Confounding 5, 7, 105, 108, 123, 131,
  133, 134, 145, 158, 161
Consumer Randomization 9, 10, 19
Conventional Randomization 23, 25, 27
Convergent Strategy 89, 90, 91, 92, 93,
  94, 102, 115
Coronary Artery Surgery Study (CASS)
  50, 143
Covariate 6, 7, 9, 11, 14, 17, 22, 23,
  100, 101, 111, 123, 124, 126, 127,
  128, 129, 130, 131, 132, 134, 143,
  144, 145, 148, 149, 150, 153, 154,
  158, 160, 162, 163, 164, 165, 166,

Selection Bias and Covariate Imbalances in Randomized Clinical Trials   V. W. Berger
© 2005 John Wiley & Sons, Ltd.

Covariate (*Continued*)
172, 173, 174, 175, 178, 179, 180,
181, 183

Deferred Enrollment 29, 143, 150,
182
Depth of Statistical Significance 76, 77,
78, 79, 80
Deterministic Allocations 86, 87, 88,
89, 90, 91, 92, 93, 94, 112, 113,
114, 115, 119, 122, 162
Dictated Allocation 20, 23
Directional Strategy 91, 92, 93, 94

Edinburgh Breast Cancer Trial 52, 53,
54, 56, 58, 60, 62
Enrollment Discretion 12, 21, 30,
130
Etanercept for Juvenile Rheumatoid
Arthritis 51
Exchangeability 4, 7
External Validity 18, 164

First-Order Residual Selection Bias 14
Forcing 20

Goteborg Mammography Trial 53, 54,
56, 58, 60, 62

Heart Outcomes Prevention Evaluation
(HOPE) Study 79, 158
Heparin for Myocardial Infarction 43
HIP Mammography Trial 53, 54, 56,
58, 60, 62
Historical Controls 5, 7
Hypertension Detection and Follow-up
Program 63

Intent-to-Randomize 160, 180, 183
Intent-to-Treat 62, 160
Internal Validity 14, 18
Ivanova-Barrier-Berger (IBB) Detection
Method 145, 146, 147
Ivanova-Barrier-Berger (IBB)
Correction Method 165

Lifestyle Heart Trial 50
Losartan Intervention for Endpoint
Reduction in Hypertension (LIFE)
Study 78

Masking 12, 13, 18, 21, 25, 26, 27, 28,
29, 39, 41, 42, 81, 107, 108, 111,
113, 121, 133, 134, 161, 162, 169,
171, 172, 174, 175, 177, 182, 183
Maximal Tolerated Imbalance (MTI)
112, 115, 116, 117, 118, 119, 121,
122, 172, 174
Matching 6, 7, 8
Maximal Procedure 101, 102, 110,
115, 116, 117, 118, 119, 120, 121,
122, 174, 175
Minimization 8, 23, 24, 28
Mistaken Allocation 43, 52, 64, 75, 77,
143

Norwegian Timolol Trial 75, 76, 77,
158
Null Hypothesis 4, 17, 24, 37, 46, 72,
124, 125, 146, 147, 167, 168

Open-Label Study 72
Oxytocin for Induction of Labor 45

Pascal Triangles 119
Phase II Studies 5
P-value 4, 8, 9, 45, 46, 50, 51, 53, 61,
63, 64, 65, 66, 67, 69, 70, 71, 72,
74, 76, 77, 78, 79, 80, 85, 125, 126,
128, 129, 147, 149, 150, 152, 159,
161
Potential Outcomes 4, 30, 85, 105,
106, 129, 147, 165, 172
Predictable Allocations 14, 21, 28, 29,
30, 35, 86, 87, 88, 89, 90, 91, 92,
93, 94, 95, 109, 111, 112, 113, 114,
115, 120, 161, 162
Primary Covariate 124, 128, 154

Random Allocation Rule 30, 122
Random Walk 111, 112, 121

Randomizable 50, 51, 143, 144, 160, 175, 176, 178, 179

Randomization as the Basis for Inference 4, 8, 9

Randomized Blocks 19, 23, 30, 45, 51, 52, 63, 80, 81, 85, 86, 87, 88, 89, 90, 91, 92, 93, 94, 96, 97, 98, 99, 100, 101, 102, 109, 110, 111, 112, 113, 114, 115, 116, 117, 118, 119, 120, 121, 122, 131, 135, 137, 138, 139, 140, 141, 142, 143, 145, 162, 163, 165, 166, 167, 174, 175, 177, 182

Randomized Response 176, 177, 178

Regression to the Mean 127

Replacement Randomization 109

Restricted Randomization 30, 72, 83, 86, 108, 109, 111, 113, 115, 116, 117, 119, 120, 122, 124, 132, 161, 177, 182, 183

Reverse Propensity Score (RPS) 124, 130, 131, 132, 133, 134, 135, 136, 137, 138, 139, 140, 141, 142, 143, 144, 145, 147, 148, 149, 150, 154, 159, 162, 163, 164, 169, 180, 181

RSV Immune Globulin for Respiratory Syncytial Virus 47

Runaway Youth Study 65, 66

Run-in Period 26, 51, 174, 175, 176, 178

Screening Log 130, 142, 143, 145, 149, 150, 154

Sealed Envelopes 43, 49, 50, 53, 63, 64, 78, 161, 179

Second-Order Residual Selection Bias 14

Selection Covariate 100, 124, 129, 130, 131, 132, 149, 153, 173, 179, 183

Stratification 20, 44, 51, 75, 81, 106, 107, 111, 133, 134, 135, 136, 137, 138, 139, 143, 149, 150, 160, 174, 182

Talc for Pleural Effusions 44

Third-Order Residual Selection Bias 14, 124, 126, 127, 128, 129, 130, 132, 133, 134, 143, 145, 149, 150, 152, 153, 154, 155

Time Trends 5, 108, 111

Tonsillectomy for Throat Infections 45

Type I Error 17, 40, 76, 85, 100, 101, 102, 115, 119, 120, 125, 147, 167, 168

Type II Error 40

University Group Diabetes Program 43

Unmasked 20, 21, 26, 27, 41, 42, 45, 47, 48, 50, 51, 52, 74, 80, 107, 108, 109, 111, 121, 122, 133, 134, 161, 165, 169, 178, 182

Unmeasured Latent Covariate 7, 11, 17, 123, 124, 129, 131, 134, 153, 158, 160, 166

Unrestricted Randomization 29, 108, 112, 122, 132, 171

Urn Randomization 22

Variative Induction 173

Varied Block Sizes 101, 102, 111, 113, 114, 115, 116, 119, 120

Western Washington Intracoronary Streptokinase Trial 46

Statistics in Practice

Human and Biological Sciences

Berger – Selection Bias and Covariate Imbalances in Randomized Clinical Trials
Brown and Prescott – Applied Mixed Models in Medicine
Ellenberg, Fleming and DeMets – Data Monitoring Committees in Clinical Trials: A Practical Perspective
Lawson, Browne and Vidal Rodeiro – Disease Mapping with WinBUGS and MLwiN
Lui – Statistical Estimation of Epidemiological Risk
Marubini and Valsecchi – Analysing Survival Data from Clinical Trials and Observation Studies
Parmigiani – Modeling in Medical Decision Making: A Bayesian Approach
Senn – Cross-over Trials in Clinical Research, Second Edition
Senn – Statistical Issues in Drug Development
Spiegelhalter, Abrams and Myles – Bayesian Approaches to Clinical Trials and Health-Care Evaluation
Whitehead – Design and Analysis of Sequential Clinical Trials, Revised Second Edition
Whitehead – Meta-Analysis of Controlled Clinical Trials

Earth and Environmental Sciences

Buck, Cavanagh and Litton – Bayesian Approach to Interpreting Archaeological Data
Glasbey and Horgan – Image Analysis for the Biological Sciences
Webster and Oliver – Geostatistics for Environmental Scientists

Industry, Commerce and Finance

Aitken – Statistics and the Evaluation of Evidence for Forensic Scientists, Second Edition
Lehtonen and Pahkinen – Practical Methods for Design and Analysis of Complex Surveys, Second Edition
Ohser and Mücklich – Statistical Analysis of Microstructures in Materials Science